Louise Adamson

WJEC CBAC

Biology AS (BY1 & BY2)

Gareth Rowlands

***WJEC** Biology AS (BY1 & BY2)*

Published by the WJEC
245 Western Avenue, Cardiff CF5 2YX

First published 2008
Second edition 2009
ISBN: 978-1-86085-643-3

Introduction

Students are often uncertain as to what is required of them in preparation for the exams

This booklet:

✓ provides guidance for students specifically studying WJEC AS Biology.
✓ is a study guide and **you should build on the information provided by using a text book** and **your lesson notes**.
✓ gives some indication of the main points to be noted.
✓ gives some guide as to the depth of treatment required.

An active role in your studies

- In order to revise effectively it is important to take an **active role** in the process. Passive reading through your notes has limited value. The guide suggests '**Action points'** where you:

✓ draw labelled diagrams.
✓ draw graphs.
✓ analyse result tables.
✓ highlight key terms and provide definitions.
✓ construct comparison tables.
✓ describe processes using flowcharts/flow diagrams.

- This does not mean that these activities are the only ones that you should carry out in preparation for an exam. They are merely suggestions. There may, for example, be a number of other drawings that you should consider.

Revision hints

Examiners are aware that candidates are often not able to effectively show the examiner what they know.

To overcome this problem it is necessary to:

✓ devise a structured **revision programme**. Revise regularly - repetition is an important tool in learning and in preparing for exams. Revise each topic as soon as it is completed. Success is achieved through consistent work rather than attempting to commit everything to memory at the last minute.
✓ **understand basic principles** e.g. by learning definitions.
✓ **revise all topics**. Each unit will contain a number of structured questions with one essay question. It is not a good idea to work out which topics are 'going to come up'. Examiners take great care that the questions cover every topic.
✓ produce **annotated diagrams** – these are a good way of revising!
✓ familiarise yourself with **past exam questions**.
✓ **express yourself clearly and concisely** -frequently students, who know a reasonable amount of biology, don't do themselves justice in an exam because of their poor exam technique. Much of this is to do with the problem of language. It can make the difference between a good grade and an average grade. You need to be able to read, interpret, memorise, understand and convey information concisely in order to answer exam questions successfully.
✓ understand the **key words** that are used in exam questions.

It is hoped that the guide may prove of real value to the student.
The author is a principal examiner for the WJEC and has written a further revision guide for the new specification 2008 together with a question book.

Revision express "A - level study guide Biology" (Pearson Education).

This publication complements the information in this booklet and provides an extension to the activities and information outlined here. This booklet, which emphasises the WJEC specific material, is largely, though not exclusively, based on the study guide.

Gareth Rowlands March 2008

Websites

www.revision-express.com

www.wjec.co.uk

1.1 Biological compounds

Certain molecules have particular functions in living organisms. These functions depend on the properties that a molecule possesses. A molecule gets its properties from its structure. You should distinguish between the terms: atom and molecule, element and compound, organic and inorganic. Water is essential as a medium for metabolic reactions; it is a constituent of cells; key elements are found in aqueous solution; it provides a habitat for aquatic organisms.

Inorganic ions

Inorganic ions play important roles in living organisms. They can be divided into two groups;

- macronutrients - needed in small amounts.

magnesium	A constituent of chlorophyll in leaves
iron	A constituent of haemoglobin in blood
phosphate	Found in plasma membrane, nucleic acids, ATP
calcium	A constituent of bones and teeth

- micronutrients - needed in minute (trace) amounts, e.g. copper, zinc

Water

- Water is **a polar** molecule and has no overall charge. The oxygen end of the molecule has a slight negative charge and the hydrogen end of the molecule has a slight positive charge. When two water molecules are in close contact their opposing charges attract each other forming a hydrogen bond. Individually the hydrogen bonds are weak but because there are many of them they stick together in a strong lattice framework. This sticking together of water molecules is called **cohesion.**
  This means that tall columns of water can be drawn up xylem vessels in tall trees.
- At ordinary temperatures water has the **highest surface tension** of any liquid except mercury. In a pond the cohesion between water molecules produces surface tension so that the body of an insect such as the pond skater is supported.
- Water is a **solvent**. Because water is a polar molecule it will attract other charged particles, such as ions, and other polar molecules, such as glucose. This allows chemical reactions to take place in solution and since these chemicals dissolve in water it acts a **transport medium** e.g. in animals blood transports many dissolved substances. In plants water transports minerals in the xylem and sucrose in the phloem. Non-polar molecules such as lipids will not dissolve in water.
  The hydrophobic property of lipids is important in cell membranes.
- Water has a high **specific heat.** A large amount of heat energy is needed to raise the temperature of water. This is because the hydrogen bonds between water molecules restrict their movement. This prevents large fluctuations in the temperature of water and this is particularly important in keeping the temperature of aquatic habitats stable so that organisms do not have to endure extremes of temperature. This also allows enzymes within the cells to work effectively.
- Water has a high **latent heat** i.e. a great deal of heat energy is needed to change it from a liquid to a vapour state. This is important, for example, in temperature control where heat is used for vapourisation of water when sweating. That is, the evaporation of water from a surface results in cooling.
- Water has a maximum density at 4°C. Water in its solid form (ice) is less dense than water and so floats on the surface. Ice forms an insulating layer and allows organisms to survive beneath it.
- Water is transparent allowing light to pass through enabling aquatic plants to photosynthesise effectively.

Carbohydrates

Carbohydrates are organic compounds containing carbon, hydrogen and oxygen.
Carbohydrates have two main **functions**:

- to act as a source of energy in both plants and animals, e.g. sugars, starch, glycogen.
- To play a structural role in plant cell walls, e.g. cellulose.

There are three main types of carbohydrates – monosaccharides, disaccharides and polysaccharides.

Monosaccharides

Monosaccharides are relatively small organic molecules and provide the building blocks for the larger carbohydrates. Monosaccharides have the general formula (CH_20) n and their name is determined by the number of carbon atoms in the molecule (n).
Where n = 3 triose - important in metabolism.
Where n = 5 pentoses - used in the formation of nucleic acids.
Where n = 6 hexoses, e.g. glucose, the main source of energy.

All sugars share the formula $C_6H_{12}0_6$ but they differ in their molecular structure. Monosaccharides usually exist as ring structures when they dissolve in water. Glucose exists as two isomers, the α form and the β form. These different forms result in considerable biological differences when they form polymers such as starch and cellulose.

❑ **Action** Draw diagrams of straight chain and ring structures of glucose.
Draw diagrams of the α and β forms of glucose.

Disaccharides

Disaccharides consist of two monosaccharide units linked together by the formation of a **glycosidic bond** *with the elimination of water*. This is called a **condensation** reaction. Disaccharides can be formed by the joining of two similar monosaccharides or by the joining of two different monosaccharides.

Monosaccharides				Disaccharide	Where found
glucose	+	glucose	=	maltose	malt sugar
glucose	+	fructose	=	sucrose	cane sugar
glucose	+	galactose	=	lactose	milk sugar

❑ ***Action***

1. Draw a molecule of maltose formed by the joining of two molecules of glucose.
2. Describe the Benedicts test for a reducing sugar such as glucose and the modified test for a non-reducing sugar such as sucrose.

Polysaccharides

Polysaccharides are large, complex molecules called **polymers.** They are formed from very large numbers of monosaccharide units linked together. The two main groups are:

- **Storage** polysaccharides such as **starch** (in plants) and **glycogen** (in animals).
  Starch is a polymer made up of many α glucose molecules held together by glycosidic bonds. It is an ideal storage molecule because:

✓ It is compact.
✓ It is insoluble and has no osmotic effect in cells.
✓ It is readily broken down to disaccharides and monosaccharides.

Starch is made up of two polymers amylose and amylopectin. Amylose is linear (unbranched) and coils into a helix whereas amylopectin is branched and fits inside the amylose.
Glycogen is sometimes called animal starch and is very similar to amylopectin.

- **Structural** polysaccharides such as **cellulose** in plants and **chitin** in insects.
  Cellulose consists of many long parallel chains of β glucose molecules cross-linked to each other by hydrogen bonds. Being made up of β glucose units, the chain has adjacent glucose molecules rotated by 180°. This allows hydrogen bonds to be formed between the hydroxyl groups of adjacent parallel chains and helps to give cellulose its structural stability. These chains are grouped together into **microfibrils.** It is these fibres that give plant cell walls their strength and rigidity.
  Chitin is similar to cellulose but has amino acids added to form a mucopolysaccharide. It is strong, waterproof and lightweight and forms the exoskeleton of insects.

Polysaccharide	**Monomer**	**Glycosidic bond**
starch (amylose)	α glucose	1-4
glycogen	α glucose	1-4 and 1-6
cellulose	β glucose	1-4

❑ ***Action*** Draw a simplified arrangement of cellulose showing cross bridges of H bonds.
Describe the iodine test for starch.

Fats(lipids)

Like carbohydrates, lipids also contain carbon, hydrogen and oxygen but in proportion to the carbon and hydrogen they contain less oxygen. They are non-polar compounds and so are insoluble in water. **Triglycerides** are formed by condensation reactions between **glycerol** and **fatty acids**. A triglyceride consists of one molecule of glycerol and three fatty acid molecules. The glycerol molecule in a lipid is always the same but the fatty acid component varies. In this reaction water is removed and an oxygen bond, known as an **ester bond**, is formed between the glycerol and fatty acid.

❑ ***Action*** Draw a representation of the following:

one molecule of glycerol + 3 molecules of fatty acids → a triglyceride + water.
(structural formula) (general formula)

There are many different fatty acids that might react with glycerol but they are of two main kinds:

- **Saturated** fatty acids where all the carbon atoms are joined by a **single** bond.
- **Unsaturated** fatty acids contain one or more **double** bonds and therefore have fewer hydrogen atoms than they might.
  A high intake of fat, notably saturated fats, is a contributory factor in heart disease.

The chemical **properties** of lipids are:

- they are insoluble in water but dissolve in organic solvents.
- fats are solid at room temperature whereas oils are liquids.

The **functions** of fats include:

- **energy storage**. Because one gram of fat yields approximately twice as much energy as one gram of carbohydrate fats are an efficient energy store. Fats are used as an energy store in seeds and animals.
- **the protection** of delicate internal organs such as kidneys.
- Triglycerides also produce a lot of metabolic water when oxidised.
  This is important in desert animals such as the camel.
- Triglycerides are also stored under the skin where they act as a heat **insulator**.

Waxes are similar to fats and oils but also contain alcohol. Waxes are insoluble and are important in land organisms such as insects where the waxy cuticle cuts down water loss. Leaves also have a waxy cuticle to reduce transpiration.

Phospholipids

Phospholipids are similar to triglycerides with one of the fatty acid groups replaced by a phosphate group.

- The lipid part is non-polar and insoluble in water (hydrophobic)
- The phosphate group is polar and dissolves in water (hydrophylic).

❑ ***Action*** Draw the structure of a phospholipid.

Phospholipids are important in the formation and functioning of membranes in cells.

Proteins

- Proteins differ from carbohydrates and lipids in that in addition to carbon, hydrogen and oxygen, they always contain nitrogen. Many proteins also contain sulphur and sometimes phosphorous.
- Proteins carry out a range of biological activities and include enzymes, antibodies, hormones, carrier and transport proteins, as well as structural proteins.
- Proteins are large compounds built up of sub-units called **amino acids.** About 20 different amino acids are used to make up proteins. There are thousands of different proteins and their shape is determined by the specific sequence of amino acids in the chain.
- All amino acids have the same basic structure in that each possesses an **amino group**, NH2-, at one end of the molecule, and a **carboxyl group**, -COOH, at the other end. It is the R group which differs from one amino acid to another.

❑ ***Action*** Draw the structural formula of a generalised amino acid.

The peptide bond

Proteins are built up from a linear sequence of amino acids. The amino group of one amino acid reacts with the carboxyl group of another with the elimination of water. The bond that is formed is called a **peptide bond** and the resulting compound is a **dipeptide**. A number of amino acids joined in this way is called a **polypeptide**.

❑ ***Action*** Draw the structural formulae of two amino acids showing the removal of one molecule of water and the formation of a dipeptide. Label the peptide link.

Protein structure

Four levels of protein structure exist:

- The **primary structure** of a protein is the sequence of amino acid in its polypeptide chain. The proteins differ from each other in the variety, numbers and orders of their constituent amino acids linked by **peptide bonds** only.
- The **secondary structure** is the shape that the polypeptide chain forms as a result of hydrogen bonding. This is most often a spiral known as the α helix. An alternative is a pleated sheet occurring as a flat zig-zag chain.
- The **tertiary structure** is formed by the bending and twisting of the polypeptide helix into a compact structure. This gives the molecule its three - dimensional shape. The shape is maintained by **disulphide, ionic** and **hydrogen bonds**. e.g. globular proteins.
- The **quaternary structure** arises from a combination of two or more polypeptide chains in tertiary form. These are associated with non-protein groups and form large, complex molecules, e.g. haemoglobin.

❑ **Action** Make sure you are familiar with the different bonds involved in each of the different levels of protein structure. What happens to the structure of enzymes when they are denatured?

Classification of proteins

Proteins can be divided into two groups according to their structure:

- **Fibrous** proteins perform structural functions. They consist of polypeptides in parallel chains or sheets with numerous cross linkages to form long fibres. e.g. keratin (in hair). Fibrous proteins are insoluble in water, strong and tough. Collagen provides tough properties needed in tendons. A single fibre consists of three polypeptide chains twisted around each other like a rope.

- **Globular** proteins perform a variety of different functions - enzymes, antibodies, plasma proteins and hormones. These proteins are compact and folded as spherical molecules. They are soluble in water. Haemoglobin consists of four folded polypeptide chains at the centre of which is an iron-containing group called haem.

❑ ***Action***

1. In a table or flowchart form make a summary of the structure, occurrence, properties and functions of proteins.
2. Construct a table comparing fibrous and globular proteins.
3. Describe the biuret test for protein.

General action points of this section.

1. You should be able to use given structural formulae (proteins, triglycerides and carbohydrates) to show how bonds are formed and broken by condensation and hydrolysis, including peptide, glycosidic and ester bonds. You should be able to recognise and understand but **not** reproduce the structural formulae of these molecules.

2. Make a list of key words and write definitions of each.

3. Make sure that you can relate the structure of carbohydrates, fats and proteins to their functions.

1.2 Cell structure

Membranes

Membranes within the cell:

- are separate areas from the rest of the cytoplasm so that potentially harmful chemicals and/or enzymes can be isolated in an organelle so that they do not damage the rest of the cell.
- provide a large surface area for the attachment of enzymes involved in metabolic processes e.g. the synthesis of ATP in the mitochondrial membranes.
- provide a transport system within the cell e.g. endoplasmic reticulum.

Cell structure

Plant and animal cells contain a number of organelles that perform a variety of functions. The detailed organisation of a cell is known as the ultrastructure of the cell.

❑ ***Action*** Draw and label a plant cell and an animal cell as seen using an electron microscope.

Draw and label each of the organelles.

(In an exam question you should be able to label a diagram of a cell and also be able to draw and label individual organelles.)

The cytoplasm is a highly organised material consisting of a soluble ground substance called the cytosol in which are found a variety of organelles:

Nucleus

This is the most prominent feature in the cell. Its function is to control the cell's activities and to retain the chromosomes. The nucleus is bounded by a double membrane, the **nuclear membrane (**or envelope**)**. This has pores in it to allow the transport of mRNA and nucleotides. The cytoplasm-like material within the nucleus is called the **nucleoplasm**. It contains **chromatin**, which is made up of coils of DNA (histones) bound to protein. During cell division the chromatin condenses to form the chromosomes. Within the nucleus is a small spherical body called a **nucleolus**. This makes RNA which is needed to make ribosomes.

Mitochondria

The mitochondrion has a double membrane separated by a narrow fluid-filled inter-membrane space. The inner membrane is folded inwards to form extensions called **cristae**. The interior of the mitochondrion contains an organic **matrix** containing numerous chemical compounds. Mitochondria are the sites of **aerobic respiration** in the cell. Some of the reactions take place in the matrix while others occur on the inner membrane. The cristae increase the surface area on which the respiratory processes take place. The function of mitochondria is to produce energy as ATP. Muscle cells contain large numbers of mitochondria reflecting the high metabolic activity taking place there.

Endoplasmic reticulum

This consists of an elaborate system of membranes forming flattened sacs. The fluid filled spaces between the membranes are called **cisternae**. The ER is connected with the nuclear membrane and may link to the golgi body. The cavities are interconnected and this system allows the transport of materials throughout the cell. There are two types of ER:

- **rough ER** has ribosomes on the outer surface. The rough ER functions in transporting proteins made by the ribosomes. Rough ER is present in large amounts in cells that make enzymes that may be secreted out of the cell.
- **smooth ER** has membranes which lack ribosomes. These are concerned with the synthesis and transport of lipids.

Ribosomes

These are made up of one large and one small sub-unit. They are manufactured in the nucleolus from ribosomal RNA and protein. They are important in **protein synthesis**.

Golgi body

This is similar in structure to ER but is more compact. The Golgi body is formed by rough ER being pinched off at the ends to form small vesicles. A number of these vesicles then join up and fuse together to form the Golgi body. Proteins are transported in the vesicles and are modified and packaged in the Golgi body. For example proteins may be combined with carbohydrates to make glycoproteins.
At the other end of the Golgi body vesicles can be pinched off and the products secreted by exocytosis when the vesicle moves to and fuses with the cell membrane.

❑ **Action** Draw a diagram showing vesicles fusing with and leaving the Golgi body.

Other functions of the Golgi body include:

- Producing secretory enzymes.
- Secreting carbohydrates e.g. for the formation of plant cell walls.
- Producing glycoprotein.
- Transporting and storing lipids.
- Forming lysosomes.

Lysosomes

Lysosomes are small vacuoles formed when portions of the Golgi body are pinched off. They contain and isolate digestive enzymes from the remainder of the cell.

- They can also release these enzymes and destroy worn out organelles in the cell. Digestion is carried out in the membrane-lined vacuole into which several lysosomes may discharge their contents.
- They can also digest material that has been taken into the cell e.g. white blood cells engulf bacteria by phagocytosis and the lysosomes discharge their contents into the vesicle so formed and digest the bacterium.

Centrioles

Centrioles are found in all animal cells and most protoctists but are absent from the cells of higher plants. Centrioles are located just outside the nucleus in a distinct region of the cytoplasm known as the centrosome. A centriole consists of two hollow cylinders positioned at right angles to one another. During **cell division** centrioles divide and move to opposite poles of the cell where they synthesise the microtubules of the spindle.

The following organelles are found in plant cells only.

Chloroplast

Chloroplasts are found in the cells of photosynthesising tissue. They have a **double membrane** inside which is the fluid **stroma,** containing ribosomes, lipid, circular DNA and possibly starch. Within the stroma are number of flattened sacs called **thylakoids.** A stack of thylakoids is called a **granum.**
Each granum consists of between two and a hundred of these closed, parallel, flattened sacs. The photosynthetic pigments such as chlorophyll are found within the thylakoids. This arrangement produces a large surface area for trapping light energy.

Vacuole

Plant cells have a large permanent vacuole which consists of a fluid-filled sac bounded by a single membrane, the tonoplast. Vacuoles contain cell sap, a storage site for chemicals such as glucose, and provide an osmotic system which functions in support of young tissues.
Animal cells contain vacuoles but these are small, temporary vesicles and may occur in large numbers.

Cellulose cell wall

The cell wall consists of cellulose microfibrils embedded in a polysaccharide matrix. The main functions of a cell wall are:

- to provide strength and support.
- to permit the movement of water from cell to cell.

❑ **Action** To aid revision draw and complete a table with the following headings:

Organelle	Simple diagram	Function

Cell organisation

There are two types of cell: prokaryotic cells and eukaryotic cells. Prokaryote cells have a simple structure and were probably the first forms of life on Earth. Eukaryotic cells probably evolved from eukaryote cells around 1000 million years ago.

Prokaryote and eukaryote cells

An example of a **Prokaryote** cell is a bacterium. **Eukaryote** cells are typical of the great majority of organisms including all animals and plants.

Prokaryotic cells	Eukaryotic cells
Bacteria and blue-green algae No membrane bound organelles DNA lies free in the cytoplasm No nuclear membrane or ER Ribosomes are smaller Cell wall containing murein	Found in plants, animals, fungi and protoctists Membrane bound organelles DNA located on chromosomes Distinct membrane-bound nucleus Ribosomes are larger Cell wall in plants made of cellulose

❑ ***Action*** Draw a labelled diagram of a bacterium.
(Labels – cell wall, membrane, plasmid, mesosome, ribosome, flagellum)

Viruses

Viruses cause a variety of infectious diseases in humans, animals and plants.
Viruses are extremely small and can only be seen using an electron microscope. They can be called 'non cells' as they have no cytoplasm, no organelles and no chromosomes. Outside a living cell a virus exists as an inert 'virion'. When they invade a cell they are able to take over the cell's metabolism and multiply within the host cell. Each virus particle is made up of a core of **nucleic acid** surrounded by a **protein coat**, the capsid. Most viruses are found in animal cells and those attacking bacteria (bacteriophages) have the nucleic acid DNA. Other animal and plant viruses contain RNA. A widely studied virus is T2 phage, a bacteriophage, which infects the bacterium *Escherichia coli* (*E.coli*)

❑ ***Action*** Draw and label a typical virus

Differences between plant and animal cells

Plant cells have all the structures found in animal cells plus some additional features:

Plant cells	**Animal cells**
cell wall	no cell wall
chloroplasts	no chloroplasts
large permanent vacuole	small, temporary vacuoles
No centriole	centriole
plasmodesmata	no plasmodesmata

❑ **Action** Construct a table to compare the structure of animal, plant and prokaryote cells and viruses.

Levels of organisation

Multicellular organisms contain a variety of different cells. As they develop, cells differentiate and become specialised to carry out specific functions e.g. muscle cells contract, nerve cells carry impulses. Some cells remain undifferentiated and function as 'packing' cells e.g. parenchyma in plants.
A **tissue** consists of a number of cells that have differentiated in the same way e.g. muscle tissue.
Different tissues combine to form an **organ** e.g. the eye.
Organs work together as **systems** e.g. excretory system.
Organisms are made up of a number of different systems that work together.

❑ **Action** microscope work

- A histological examination of the following animal cells:
- ✓ cuboidal and ciliated epithelium.
- ✓ smooth and striated muscle.
- ✓ collagen connective tissue.
- Look at a range of living cells such as onion epidermis, Canadian pondweed (*Elodea*) cells, potato tuber cells, *Spirogyra.*
- Study a range of electron micrographs of prokaryote and eukaryote cells.

1.3 Cell membrane

Cell membrane

The cell surface membrane or plasma membrane is the boundary that separates the living cell from its non-living surroundings. The membrane also controls which substances pass into and out of the cell.

Structure of the membrane

The cell membrane is made up almost entirely of **proteins** and **phospholipids**.

The phosphate head of the phospholipid is a polar molecule (hydrophilic or water-loving) and has an attraction for other polar molecules such as water. The fatty acid end of the phospholipid, made up of two fatty acid tails is non-polar (hydrophobic or water-hating) and repels water. Phospholipids can form bilayers with one sheet of phospholpid forming over another. This phospholipid bilayer forms the basis of membrane structure. The use of the electron microscope enabled Singer and Nicholson to put forward their theory called the **Fluid Mosaic model** in 1972. They proposed that:

- there is a **bimolecular phospholipid** layer.
- associated with the bilayer is a variety of protein molecules:
- ✓ some of the proteins occur on or in only one of the layers (**extrinsic** proteins)
- ✓ some proteins extend across both layers (**intrinsic** proteins)
- the phospholipid layer is capable of movement i.e. it is fluid and in surface view the proteins are dotted throughout the layer in a mosaic arrangement.
- The model is referred to as the 'fluid mosaic' model because the components are free to move with respect to each other.

❑ ***Action*** Draw a diagram showing the fluid mosaic structure of the cell membrane.

Cholesterol is also found in animal cells. It fits between the phospholipid molecules, increasing the rigity and stability of the membrane. Glycolipids (lipids that have combined with polysaccharide) are also found in the outer layer of the membrane and are thought to be involved in cell-to-cell recognition. Glycoproteins also stick out of some membranes.

The main **functions** of the cell membrane include:

- structural integrity
- secreting chemicals
- cell –cell recognition
- taking up nutrients and other requirements.

The membrane as a barrier

The cell surface membrane is selectively permeable to water and some solutes. Lipid-soluble substances can move through the cell membrane more easily than water-soluble substances.

- Small, uncharged molecules, such as oxygen and carbon dioxide, freely pass through the membrane.
- Lipid soluble molecules such as glycerol can pass through the membrane.
- The hydrophobic core of the membrane impedes the transport of ions and polar molecules.

Charged particles (ions) and relatively large molecules such as glucose cannot diffuse across the non-polar centre of the phospholipid bilayer because they are relatively insoluble in lipid. Intrinsic proteins assist such particles to pass in or out of the cell by a passive process called **facilitated diffusion**. (These proteins are of two types: channel proteins and carrier proteins, and are dealt with in more detail in the next section).

Transport across cell membranes

Diffusion

Diffusion is the movement of molecules or ions from a region where they are in high concentration to a region of lower concentration until they are equally distributed. Ions and molecules are always in a state of **random movement** but if they are highly concentrated in one area there will be a net movement away from that area until equilibrium is reached or until there is a uniform distribution.
The rate of diffusion is affected by:

- the concentration gradient i.e. the greater the difference in the concentration of molecules in two areas, the greater the rate.
- the distance of travel i.e. the shorter the distance between two areas the greater the rate.
- the surface area of the membrane -the larger the area the quicker the rate.
- the thickness of the membrane - the thinner the membrane the greater the rate.
- and an increase in temperature results in an increase in rate since there is an increase in molecular energy and therefore movement.

Facilitated diffusion

Charged particles or ions and large molecules such as glucose do not readily pass through the cell membrane because they are relatively insoluble in lipid. In the cell membrane protein molecules span the membrane from one side to the other and help such particles to diffuse in or out of the cells. These proteins are of two types:
Channel proteins - consist of pores lined with polar groups allowing charged ions to pass through. (As the channel is hydrophilic, water-soluble substances can pass through.) As each channel protein is specific for one type of ion each protein will only let one particular ion through. They can also open and close according to the needs of the cell.

Carrier proteins allow the diffusion across the membrane of larger polar molecules such as sugars and amino acids. A particular molecule attaches to the carrier protein at its binding site and causes the carrier protein to change its shape, releasing the molecule through the membrane.

Carrier proteins and channel proteins increase the rate of diffusion along the concentration gradient without the need for energy in the form of ATP from respiration.

❑ **Action** Draw diagrams to show how carrier and channel proteins operate in the membrane.

Osmosis

Most cell membranes are permeable to water and certain solutes only. In biological systems osmosis is a special form of diffusion which involves the movement of **water** molecules only**.**
Osmosis may be defined as the movement of water from a region where it is highly concentrated to a region where it is lower, through a partially permeable membrane.
Biologists use the term **water potential (WP)** ψ(psi) to describe the tendency of water molecules to move from high to low concentrations.
Pure water has the highest water potential of zero. This is because where there is a high concentration of water molecules they have a greater potential energy i.e. the water molecules are completely free to move about. When a solute, such as sugar, is dissolved in water there are proportionally fewer water molecules to move about and the water potential of the solution is lowered. All water potentials (except that of pure water) have a negative value. The more concentrated the solution the more negative the water potential i.e. the fewer free water molecules there are. We can now define osmosis in terms of water potential.

Osmosis is the diffusion of water molecules from a region of high water potential to a region of lower water potential through a partially permeable membrane.

A higher WP implies a greater tendency of water to leave a system. Water will diffuse from a region of less negative (higher) WP to one of more negative (lower) WP.

In plant cells the following equation is used to describe the relationship between the forces:

$$\underset{\text{water potential}}{\psi} = \underset{\text{solute potential}}{\psi_s} + \underset{\text{pressure potential}}{\psi_p}$$

- The presence of solute molecules in the vacuole of a plant cell lowers the WP.
- The concentration of dissolved substances inside the cell vacuole is called the **solute potential.**
- When water enters a plant cell vacuole by osmosis a hydrostatic pressure is set up and pushes outwards on the cell wall. As the outward pressure builds up the cell wall develops an opposing force called the **pressure potential.** The pressure potential is usually positive.

Turgor and plasmolysis

- If the WP of the external solution is lower than the solution inside the cell it is said to be **hypertonic** and water flows out of the cell.
- If the WP of the external solution is higher than the solution inside the cell it is said to be **hypotonic** and water flows into the cell.
- If the cell has the same solute concentration as the surrounding solution the external solution is **isotonic** with that of the cell.

- When a plant cell is placed in a hypertonic solution it loses water by osmosis. The vacuole shrinks and the cytoplasm will draw away from the cell wall. This process is called **plasmolysis** and when complete the cell is said to be **flaccid.**
- The point at which the cell membrane *just* begins to move away from the cell wall is said to be the point of **incipient plasmolysis.**
- A plant cell will gain water if placed in a hypotonic solution and will continue to take in water until prevented by the opposing wall pressure. The pressure potential rises until it is equal and opposite to the solute potential. In theory the water potential is now zero and when the cell can not take in any more water it is said to be **turgid**. The state of **Turgor** is important in plants, particularly young seedlings. It supports them and maintains their shape and form.
- An animal cell will burst if placed in a hypotonic solution as it has no cell wall.

❑ ***Action***

1. Draw labelled diagrams of a flaccid and a turgid plant cell.
2. Make a list of key terms and their definitions.
3. Be prepared to use the given equation in calculations.

Endocytosis and exocytosis

These are processes where the cell transports materials in bulk.

- **Endocytosis** – the cell surrounds the membrane around the material and brings it into the cytoplasm inside a vesicle. There are two types of endocytosis:

1. **Phagocytosis** is the process by which the cell can obtain **solid** materials that are too large to be taken in by diffusion or active transport. A lysosome fuses with the vesicle formed, enzymes digest the solid material and the products are absorbed into the cytoplasm. Phagocytes (white blood cells) destroy bacteria and remove cell debris by phagocytosis.
2. **Pinocytosis** is entry of **liquid** by the same mechanism as phagocytosis, except that the vesicles produced are smaller.

- **Exocytosis** refers to substances leaving the cell after being transported through the cytoplasm in a vesicle. Digestive enzymes are often secreted in this way.

Through phagocytosis and secretion the cell membrane is continually having portions removed or added to it.

❑ ***Action*** Draw a diagram to illustrate phagocytosis.

Active transport

Unlike the processes described so far, active transport is an **energy** requiring process in which **ions** and **molecules** are **moved** across membranes **against a concentration gradient.**

The features of active transport are:

- Ions and molecules can move in the opposite direction to that in which diffusion occurs, i.e against a concentration gradient.
- The energy for active transport is supplied by ATP, and anything which affects the respiratory process will affect active transport.
- Active transport will not take place in the presence of a respiratory inhibitor such as cyanide.
- The process occurs through the carrier proteins that span the membrane. The proteins accept the molecule and then the molecule enters the cell by a change in shape of the carrier molecule.

Processes involving active transport include protein synthesis, muscle contraction, nerve impulse transmission, absorption of mineral salts by plant roots.

❑ ***Action*** Draw diagrams to show how the carrier proteins change shape when conveying a molecule across the membrane.

1.4 Enzymes

In cells metabolic reactions take place quickly and thousands of reactions are taking place simultaneously. Order and control is essential if reactions are not to interfere with each other. These features of metabolism are made possible by the action of enzymes.

Structure of enzymes

All enzymes are tertiary globular proteins where the protein chain is folded back on itself into a spherical or globular shape. Each enzyme has its own sequence of amino acids and is held in its tertiary form by hydrogen bonds, disulphide bridges and ionic bonds. This complex three-dimensional shape gives the enzyme many of its properties. (Review earlier work on the structure of proteins.)

How enzymes work

Enzymes are **biological catalysts** that speed up the rate of metabolic reactions.
These reactions can be of two types:

1. Reactions where larger molecules are broken down into smaller molecules.
2. Reactions where small molecules are built up into larger, more complex, molecules.

Enzymes react with another molecule called a **substrate.** Each enzyme has its own special shape, with an area, the **active site**, onto which the substrate molecules bind.
Modern interpretations of the lock and key theory suggest that in the presence of the substrate the active site may change in order to select the substrate's shape.
This is called the **induced fit hypothesis**.
(This may be illustrated by the enzyme, lysozyme, but structural details are not required.)

Enzyme + substrate = enzyme- substrate complex = enzyme + product.

Properties of enzymes

- Enzymes are **specific** i.e. each enzyme will catalyse only one particular reaction, for example, sucr**ase** acts on the sugar, sucr**ose**.
- Enzymes are very efficient and have a high **turnover number.** This means that they can convert many molecules of substrate per unit time, for example, catalase, which breaks down the waste product hydrogen peroxide in the body, has a turnover number of several million!
- Chemical reactions need energy to start them off and this is called **activation energy**. This energy is needed to break the existing chemical bonds inside molecules. In the body enzymes lower the activation energy of a reaction and so reduce the input of energy needed and allow reactions to take place at lower temperatures.

Factors affecting the rate of enzyme action

Enzymes are made inside living cells but may act inside the cell (intracellular) or outside (intercellular, extracellular) such as the digestive enzymes of the alimentary canal. Environmental conditions, such as temperature and pH, change the three dimensional structure of enzyme molecules. Bonds are broken and hence the configuration of the active site is altered.

- **Temperature**

An increase in temperature gives molecules greater kinetic energy and they move around more quickly increasing the chance of molecules colliding. Increasing the temperature of an enzyme controlled reaction results in an increase in the rate of reaction. As a general rule, the rate of reaction doubles for each 10°C rise in temperature until an optimum temperature is reached. For most enzymes this is 40°C. Above this temperature the increasing vibration of the molecules cause the hydrogen bonds to break and cause a change in the tertiary structure of the enzyme. This alters the shape of the active site.

and the substrate molecule will no longer fit. The enzyme is then said to be **denatured**. This is a permanent change in the structure. If enzymes are subjected to low temperatures, such as freezing the enzyme is inactivated as the molecules have no kinetic energy. However, the enzyme can work again if the temperature is raised.

- **pH**

The rate of an enzyme catalysed reaction will vary with changes in pH. Enzymes have a narrow optimum range and small changes in pH can affect the rate of reaction without affecting the structure of the enzyme. Small changes in pH outside the optimum can cause small reversible changes in enzyme structure and results in inactivation. Extremes of pH can denature an enzyme.
The charges on the amino acid side-chains of the enzyme's active site are affected by free hydrogen ions or hydroxyl ions. In the formation of an enzyme substrate complex the charge on the active site must match those on the substrate. If the active site has too many $H+$ ions (say) the active site and the substrate may both have the same charge and the enzyme will repel the substrate.
At extremes of pH the hydrogen bonding is affected and the three-dimensional shape of the enzyme is altered and so is the shape of the active site.

Enzymes are also affected by the concentration of the substrate and the concentration of the enzyme itself.

- **Substrate concentration**

The rate of an enzyme catalysed reaction will vary with changes in substrate concentration. If the amount of enzyme is constant the rate of reaction will increase as the substrate increases. But there must come a point when all the enzyme's active sites are working to full capacity i.e. all the active sites are filled.

- **Enzyme concentration**

The rate of an enzyme catalysed reaction will vary with changes in enzyme concentration.
Increasing the enzyme concentration will increase the rate of reaction.

Action

1. Make a list of key terms and give a definition for each term.
2. Draw separate graphs to illustrate the effect of the above factors on enzyme action.
3. Describe, **in detail**, the shape of the graph in each case.

Enzyme experiments

In enzyme experiments it is essential that buffers and controls are used.

✓ **Buffers** maintain a constant pH. When a buffer is used in an experiment the pH changes little when a small quantity of acid or alkali is added. It may be said that a buffer 'soaks up hydrogen ions'.

✓ **Control** – controls are duplicate experiments, identical in every respect to the actual experiment, except for the variable being investigated, which is kept constant. For example, boiled enzyme may be used in a control experiment instead of the enzyme.

Enzymes and inhibitors

Inhibition occurs when enzyme action is slowed down or stopped by another substance. The inhibitor combines with the enzyme and stops it forming an enzyme-substrate complex.
There are two types of inhibitors:

- **Competitive**, where the inhibitor is **structurally similar to the substrate** and competes with the active site of the enzyme, i.e. the inhibitor has a shape that lets it fit into the active site of the enzyme in place of the substrate. For example, malonic acid competes with succinate for the active sites of succinic dehydrogenase, an important enzyme in the Krebs cycle in respiration. If the substrate concentration is increased it will reduce the effect of the inhibitor. This is because the more substrate molecules present the greater the chance of finding active sites, leaving fewer to be occupied by the inhibitor.

- **Non-competitive**, where the inhibitor **binds to the enzyme at a site away from the active site**. This alters the overall shape of the enzyme molecule, including the active site, in such a way that the active site can no longer accommodate the substrate. As the substrate and inhibitor molecules attach to different parts of the enzyme they are not competing for the same sites. The rate of reaction is therefore unaffected by substrate concentration.
  For example, cyanide (a respiratory poison) attaches itself to part of the enzyme, cytochrome oxidase, and inhibits respiration.

❑ **Action** Draw diagrams to illustrate the different types of inhibition.
Draw a graph to show how the two types of inhibition affect the rate of reaction of an enzyme when the substrate concentration is increased.

1.5 Medical and industrial applications of enzymes

Enzymes are used on a wide commercial scale in the food, pharmaceutical and agrochemical industries

Immobilised enzymes are enzyme molecules that are fixed, bound or trapped on an inert matrix such as a gel capsule (alginate beads). These beads can be packed into glass columns. Substrate can be added to the top of the column and it reacts with the enzyme as it slowly flows down the column. Once set up the column can be used again and again. As the enzyme is fixed it does not get mixed up with the products and is therefore cheaper to separate. Immobilised enzymes are used widely in industrial processes, such as fermentation, as they can readily be recovered for reuse.

Advantages of immobilised enzymes.

Enzyme instability is one of the key factors that prevent the wider use of 'free' enzymes. Chemicals such as organic solvents, raised temperatures and pH values outside the norm can denature the enzyme with a consequent loss of activity. Immobilising enzymes with a polymer matrix creates a microenvironment. By allowing processes to occur at higher temperatures than normal means that activity is increased and so production is increased. Other advantages include:

- ✓ Enzymes can **tolerate** a wider range of conditions.
- ✓ Enzymes are **easily recovered** for re-use thus reducing overall costs.
- ✓ **Several enzymes** with differing pH or temperature optima **can be used together.**
- ✓ Enzymes can be **easily added or removed** giving greater control over the reaction.

One use of immobilised enzymes involves **biosensors,** which work on the principle that enzymes are specific and are able to select one type of molecule from a mixture even in very low concentrations. A biosensor can be used in the rapid and accurate detection of minute traces of biologically important molecules.

Biosensors have great potential in the areas of medical diagnosis and environmental monitoring. The electrode probe can detect changes in substrate or product, temperature changes or optical properties. One particular use of a biosensor is in the **detection of blood sugar** in diabetics. Here the biosensor uses an enzyme, together with a transducer, which produces an electrical signal in response to substrate transformation. The strength of the electrical signal may be measured with a suitable meter.

The electrode probe, which has a specific enzyme immobilised in a membrane, is placed in the blood sample. If glucose is present it diffuses through the membrane, forms an enzyme-substrate complex. The reaction produces a small electric current, which is picked up by the electrode (the transducer). This current is read by a meter which produces a reading for blood glucose. Normal blood glucose levels are described as being 3.89 – 5.83 mmol dm^{-3}.

Steps in using a biosensor.

1. Blood contains a mixture of different molecules.
2. Enzyme electrode is placed in a blood sample.
3. Glucose diffuses into the immobilised enzyme layer.
4. Oxygen is taken up.
5. The rate of oxygen uptake is proportional to the glucose concentration.
6. A digital display shows an accurate concentration of glucose.

Immobilised enzymes are also used in fermenters. Immobilised enzymes provide a rapid, sensitive and specific measurement of products.

❑ ***Action*** 1. Draw and label a biosensor.
2. Construct a table of comparison between 'free' and immobilised enzymes.

1.6 Nucleic acids

The genome is all the DNA sequences contained in the chromosomes of an organism.

Nucleic acids

There are two types of nucleic acid: deoxyribonucleic acid (DNA) and ribonucleic acid (RNA).
Both are built up of units called **nucleotides**.
Individual nucleotides are made up of **three parts** that combine by condensation reactions. These are:

- **phosphoric acid** (phosphate H_3PO_4). This has the same structure in all nucleotides.
- **pentose sugar**, of which there are two types:
  - ✓ in ribonucleic acid (RNA) the sugar is ribose.
  - ✓ in deoxyribonucleic acid (DNA) the sugar is deoxyribose.
- **organic base** which contains nitrogen.

There are five different bases which are divided into two groups:

- the **pyrimidine bases (single ring structures) are thymine, cytosine and uracil.**
- the **purine** bases (double ring structures) are adenine and guanine.

Deoxyribonucleic acid (DNA)

The structure of DNA

- DNA is a double stranded polymer of nucleotides or **polynucleotide**.
  (A polymer is a large number of repeating units).
- Each polynucleotide may contain many million nucleotide units.
- It is in the form of a **double helix**, the shape of which is maintained by hydrogen bonding.
- The pentose sugar is always **deoxyribose.**
- DNA contains four **organic bases**. These are adenine, guanine, cytosine, and thymine.
- Each strand is linked to the other by pairs of organic bases.
- Cytosine always pairs with guanine, adenine always pairs with thymine, and the bases are joined by hydrogen bonds.
- Put simply, DNA is like a coiled ladder with the uprights of the ladder being made up of alternating sugars and phosphate groups and the rungs made up of the bases. The bases are held together by weak hydrogen bonds.

❑ ***Action*** Draw a part of a DNA chain showing a polynucleotide consisting of three base pairs.

DNA performs two major functions:

1. **Replication** in dividing cells.
2. Carrying the information for **protein synthesis**.

Ribonucleic acid (RNA)

The structure of RNA.

- RNA is a **single stranded** polymer of nucleotide.
- RNA contains the pentose sugar, **ribose.**
- RNA contains the organic **bases** adenine, guanine, cytosine, and uracil (in place of thymine).

There are three types of RNA and all are involved in the process of protein synthesis.

- **messenger RNA (mRNA)** is a long single-stranded molecule formed into a helix. It is manufactured in the nucleus and carries the genetic code from the DNA to the ribosomes in the cytoplasm.

- **ribosomal RNA (rRNA)** is found in the cytoplasm and is a large, complex molecule made up of both double and single helices. Ribosomes are made up of ribosomal RNA and protein.

- **transfer RNA (tRNA)** is a small single stranded molecule. It forms a clover-leaf shape, with one end of the chain ending in a cytosine-cytosine-adenine sequence at which point the amino acid it carries attaches itself. At the opposite end of the chain is a sequence of three bases called the **anticodon.** tRNA molecules bring amino acids to the ribosome so that proteins can be synthesised.

❑ ***Action***

1. Draw the different types of RNA.
2. Construct a table of differences between DNA and RNA.

The process of protein synthesis is studied in A2 but a simplified version has been included in this section **for information only** so as to gain an understanding of the functions of the different forms of DNA in the process of protein synthesis.

- DNA carries all the information to make a protein but it cannot leave the nucleus.
- Protein synthesis takes place on the ribosomes that are found in the cytoplasm.
- So there must be a messenger molecule to carry the coded information from the DNA through the nuclear membrane to the cytoplasm. This single stranded molecule is **mRNA.**
- Having left the nucleus the mRNA attaches itself to a **ribosome.**
- Here the code is translated as the ribosome moves along the mRNA and exposes three bases (codon) at a time.
- The role of the tRNA molecules is to carry specific amino acids to the ribosomes.
- The anticodon on the **tRNA** joins with the codon of the mRNA and the amino acid attaches itself to the ribosome.
- This process is repeated as the ribosome moves along the mRNA each time leaving an amino acid attached to the previous amino acid until a polypeptide chain is formed.

1.7 Cell division

Chromosome structure

Chromosomes are made up of **DNA**, protein and a small amount of RNA. DNA occurs as a single strand in the form of a double helix running the length of the chromosome. Each DNA molecule is made up of many small sections called **genes**. It is only at the onset of cell division that chromosomes become visible. Shortly before cell division begins each DNA molecule makes a copy of itself. The single thread of DNA becomes two identical threads. These are called **chromatids** and they lie parallel along most of their length but are joined only in a specialised region called the **centromere**.

The number of chromosomes in the cells of different species varies. Humans always have 46 chromosomes, a fruit fly has 8 chromosomes, a potato 48 chromosomes! Chromosomes are found in matching pairs, called **homologous** pairs. So humans have 23 pairs of homologous chromosomes. The total number of chromosomes is called the diploid number. Sex cells, or gametes, have half the diploid number, this is called **haploid.** So human gametes have 23 chromosomes.

Mitosis

Mitosis produces two daughter cells that are genetically identical to the parent cell.
Dividing cells undergo a regular pattern of events known as the **cell cycle**. This is a continuous process but for convenience of description it is subdivided into four stages plus a 'resting' stage, known as interphase, between one complete division.

Interphase
This is the longest part of the cycle during which a newly formed cell increases in size and **produces organelles** lost during the previous division. The amount of **DNA is doubled** during this period. Just before the next cell division the chromosomes replicate so that each then consists of two chromatids joined together by the centromere. There is considerable metabolic activity as these processes need energy in the form of ATP. The chromosomes are not visible at interphase because the chromosome material, chromatin, is dispersed throughout the nucleus.

During **prophase** the following changes take place:

1. The chromosomes condense (shorten and become thicker) and become visible as long thin threads. They are now referred to as pairs of chromatids.

2. In cells where centrioles are present i.e. animals and lower plants, the centrioles, which have divided, move to the opposite ends (poles) of the cells.

3. Protein microtubules form from each centriole and the spindle develops, extending from pole to pole.

4. Towards the end of prophase the nuclear membrane disintegrates and the nucleolus disappears.

5. Pairs of chromatids can clearly be seen lying free in the cytoplasm.

During **Metaphase** the chromosomes arrange themselves at the centre or equator of the spindle and become attached to certain spindle fibres at the centromere. Contraction of these fibres draws the individual chromatids slightly apart.

Anaphase is a very rapid stage**.** The centromere splits and the spindle fibres contract and pull the now separated chromatids to the poles, centromere first.

Telophase is the final stage of mitosis. The chromosomes have now reached the poles of the cells and are referred to as chromosomes again. They uncoil and lengthen. The spindle breaks down, the centrioles replicate, the nucleoli reappear and the nuclear membrane reforms. In animal cells cytokinesis occurs by the constriction of the centre of the parent cell from the outside inwards. In plant cells, a cell plate forms across the equator of the parent cell from the centre outwards and a new cell wall is laid down.

❑ ***Action***

1. Make a list of the key words and give their definitions. e.g. **homologous** means that in the diploid cell each chromosome has a partner of exactly the same length and with precisely the same genes.
2. Construct a table summarising the changes that take place at each of the four stages of mitosis.
3. Construct a pie chart showing the relative length of time of each period of the cell cycle.
4. Draw labelled diagrams showing the stages in mitosis.

Significance of mitosis

- Mitosis produces two cells that have the same number of chromosomes as the parent cell and each **chromosome is an exact replica** of one of the originals.The division allows the production of cells that are **genetically identical to the parent** and so gives genetic stability.
- By producing new cells, mitosis leads to **growth** of an organism and also allows for **repair** of tissues and the replacement of dead cells. An example of mitosis in plants is in the root tip. In human skin, dead surface cells are replaced by identical cells from below.
- **Asexual reproduction** results in complete offspring that are identical to the parent.
  This takes place in unicellular organisms such as yeast and bacteria. It also takes place in certain flowering plants where organs such as bulbs, tubers and runners produce large numbers of identical offspring in a relatively short period of time. There is no variation between each individual. However, most of these plants also reproduce sexually.

Meiosis

- It is important that gametes are haploid since at fertilisation half the chromosomes come from the male and half from the female in order to produce a diploid zygote.
- Meiosis produces four genetically different haploid cells during sexual reproduction and involves two consecutive divisions.

Meiosis I - where the chromosome number is reduced and crossing over takes place.
Meiosis II - where the two new haploid nuclei divide again in a division identical to that of mitosis.

Like mitosis, meiosis is a gradual process but for convenience it is divided into the four phases of prophase, metaphase, anaphase and telophase, these phases occurring once in each of the two divisions.

*You are **not** required to describe the complete process of meiosis but to describe the significance of the **differences** between mitosis and meiosis.*

The first stage of **prophase** in meiosis is similar to prophase in mitosis in that the chromosomes shorten and fatten and become visible but in meiosis they **associate in their homologous pairs**. The pairing of the chromosomes is called **synapsis**. Each pair of chromosomes, one from the mother and one from the father, is called a **bivalent**. Each bivalent consists of four strands made up of four chromosomes each divided into two chromatids. These chromatids wrap around each other and then partially repel each other but remain joined at certain points called **chiasmata**. At these points chromatids may break and recombine with a different but equivalent chromatid. This swapping of pieces of chromosomes is called **crossing over** and is a source of genetic variation.

During the first stage of **metaphase** of meiosis the pairs of homologous chromosomes arrange themselves randomly on the equator of the spindle. Chance determines how the homologous chromosomes are arranged on the equator and when they separate different combinations are produced. This **random distribution** and consequent i**ndependent assortment** of chromosomes produces new genetic combinations.

Table comparing mitosis and meiosis.

Mitosis	Meiosis
One division resulting in two daughter cells	Two divisions resulting in four daughter cells
Number of chromosomes is unchanged	Number of chromosomes is halved
Homologous chromosomes do not associate in pairs	Homologous chromosomes pair up
Crossing over does not occur	Crossing over occurs and chiasmata form
Daughter cells are genetically identical	Daughter cells are genetically different

Meiosis and variation

In the long term, if a species is to survive in a constantly changing environment and to colonise new environments, sources of variation are essential. There are three ways of creating variety:

- Each of the chromosomes making up a homologous pair carries different genetic material. During sexual reproduction the **genotype of one parent is mixed with that of the other** when haploid gametes fuse.
- The pairs of **homologous chromosomes** arrange themselves randomly on the spindle during metaphase 1 of meiosis. When they subsequently **separate** they do so entirely i**ndependently** of each other, so that the daughter cells or gametes contain different combinations of chromosomes.
- **Crossing over** during chiasmata formation during prophase 1 of meiosis. Equivalent parts of homologous chromosomes may be exchanged thus producing new combinations and the separation of linked genes.

2.1 Bio-diversity and Evolution

Bio-diversity is a measure of the number of species on the planet; extinction is the loss of species. Over the last 200 years certain human activities have had negative effects on the environment and this has, in turn, affected the survival of plants and animals. Now scientists have come to realise that there is a bio-diversity crisis, a rapid decrease in the variety of life on Earth. Ecosystems such as tropical rain forests and coral reefs are the most diverse habitats on Earth and have some of the greatest concentrations of species per square kilometre compared to other ecosystems. Yet, tropical rain forests are being destroyed at an alarming rate to make room for, and to support, the increase in the human population. Human alteration to habitat is the single greatest threat to bio-diversity on the planet.

Bio-diversity and extinction

Human activities are altering ecosystems upon which they and other species depend. In the oceans stocks of many fishes are being depleted by over-harvesting, and some of the most productive and diverse areas, such as coral reefs and estuaries are being severely stressed. Globally, the rate of species loss may be as much as 50 times higher than at any time in the past 100,000 years.
Extinction is a natural process that has been taking place since life first evolved. It is the current ***rate*** of extinction that underlies the bio-diversity crisis. Scientists believe that the normal 'background' rate of extinction is one out of every million species per year. It is now estimated that human activity in tropical areas alone has increased extinction rates between 1000 and 10,000 times! Massive destruction of habitats throughout the world have been brought about by agriculture, urban development, forestry, mining, and environmental pollution. Marine life has also been affected. About one third of the planet's marine fish species rely on coral reefs. At the current rate of destruction about half of the reefs could be lost in the next 20 years.

Endangered species

The vast majority of Earth's earlier occupants, including the large and once dominant dinosaurs and tree ferns, have become extinct largely as a result of climatic, geological and biotic changes. At the present time, human activity has taken over as the main cause of species evolution. Many of the larger mammals e.g. mountain gorillas, giant pandas, tigers and polar bears are threatened.
Their decline in numbers has three main causes:

- loss of habitat
- over-hunting by humans
- competition from introduced species.

Other species are also threatened by additional causes such as:
- deforestation
- pollution
- drainage of wetlands.

It is now recognised that each species may represent an important human asset, a potential source of food, useful chemicals, or disease-resistant genes. For example, of the many plants growing in the tropical rain forests there may be some with medicinal properties. The extinction of any plant species before their chemical properties have been investigated could amount to an incalculable loss. There is therefore a need for **species conservation**, the planned preservation of wildlife.

Evolution

What has brought about the existence of so many different life-forms on Earth?
Evolution is the process by which new species are formed from pre-existing ones over very long periods of time. In biology the term is used specifically for the processes that have transformed life on Earth from its early beginnings to the vast diversity of fossilised and living forms that are known today.
The theory of evolution was first put forward by Charles Darwin. During his visit to the Galapagos Islands Darwin accumulated geological and fossil evidence that supported the idea that life changes with time. In 1859 he proposed natural selection as the force that causes changes in populations.

Adaptive radiation in the finches on the Galapagos Islands

In 1832, Darwin, then aged 22, travelled to South America to carry out a scientific survey. He studied the flora and fauna of mainland South America and of some surrounding islands, including the Galapagos Islands. These islands were formed in recent geological time, the result of volcanic activity. His task was to observe, describe and classify the plants and animals that he found. He also collected fossils in the rocks and these showed him that different lifeforms had gone through many changes in the past.

Among the many animals that Darwin studied on the Galapagos Islands were the finches of which he observed fourteen different species. In geological terms the islands were recently formed and any animals found there must have reached the islands from the mainland, some 600 miles away. Finches are unable to fly long distances and Darwin suggested that one ancestral species of finch had reached the islands with the help of the prevailing winds. As there were no other bird species inhabiting the islands there was a variety of food available to the colonising finches. He noticed how individual finches differed from one island to the next. The main differences were in the size and shape of their beaks and these were related to the different type of food eaten, e.g. insects, seeds, fruit.

❑ **Action** Draw the different types of finch beaks and describe how the beaks are adapted to different diets.

It seemed that on each island the characteristics that best suited a particular finch to its environment were inherited by the offspring. Darwin suggested that the finches had developed from a common ancestor and that the type of beak had developed over time and become specialised to feed on a particular food source. This is an example of **adaptive radiation.**

Natural selection

Darwin's observations of variation within a population led him to propose the idea of natural selection.

- ✓ There is a struggle for survival (**competition**).
- ✓ The individuals that **survive and reproduce** pass on to their offspring favourable characteristics.
- ✓ In time, a group of individuals that once belonged to the same species may give rise to two different groups that are sufficiently distinct to belong to two separate species.

A species may be defined as a group of organisms which share a large number of common characteristics and which can interbreed to produce fertile offspring. If a group of organisms becomes separated into two more isolated populations, these may become sufficiently different to prevent members of one group breeding with individuals from the other group. They have become two distinct species.

Gradual change over long periods of time or very sudden leaps?

There is a great deal of evidence to suggest that the Earth's climate changed in the past. About three million years ago the Earth's climate was much warmer than it is now. Then forests grew as far north as the Arctic coast of Greenland. The ice caps covering the poles were thinner, and sea levels were about 35 metres higher than they are today. The first ice age commenced around 2.3 million years ago. How do scientists know about these changes and how have these changes affected life on the planet?

Palaeontology is the study of plants and animals of the geological past, as represented by their fossil remains. By arranging extinct animals and plants into geological sequence, it is possible to suggest how one group may have evolved into another. The fact that fossils are formed in sedimentary rocks helps palaeontologists to do this. In the formation of sedimentary rocks, layers of silt harden, and form layers on top of each other. The resulting rock consists of a series of horizontal layers (strata) with each layer containing fossils typical of the time when it was laid down. The oldest rocks, and therefore the earliest fossils are contained in the lowest layers; the youngest rocks, and the most recent fossils, are in the uppermost layers. Rocks can often be dated precisely by radiometric dating techniques. Fossils are dated by the radiocarbon method. Knowing the age of the rocks and a study of the fossil record tells scientists about the sequence and timing of the appearance of the major groups of living organisms.

Darwin's theory is based on the idea that species gradually change over long periods of time from one form to another. Until recently almost all biologists took a similar viewpoint. If this idea is correct, it would be expected that intermediate forms, or 'missing links', would be found in successive rock layers between one fossil species and the next. However, intermediate forms in the fossil record are surprisingly rare.

- Creationists believe that this rarity of intermediate forms is evidence of 'special creation' rather than in the evolution of species.
- Two American scientists, Eldridge and Gould have put forward a different interpretation. They suggest that new species may arise rapidly, perhaps within a few thousand years, and then remain unchanged for millions of years before changing again. This sudden evolution of a new species may take place at the edges of the area that the population inhabits and that only a *few* individuals are involved. Under these circumstances, it would be unusual to find a gradation between successive species in the fossil record. When the environment changes unfavourably it causes species to move to better conditions. So organisms migrate as well as evolve! Those that are unable to move to a favourable habitat may die out at this stage. The present global warming effect is causing the polar ice caps to melt and affect the organisms that inhabit the area. Polar bears are in danger of becoming extinct.

  When the changes are extreme mass extinctions occur.

Mass extinctions

Scientists believe that there have been a number of mass extinctions, suggesting that perhaps there hasn't been a 'gradual change' as predicted by the theory of natural selection. For example, there was a mass extinction 250 million years ago where over 80% of marine invertebrate groups were eliminated! Another mass extinction occurred 65 million years when the dinosaurs were wiped out. In a great wave of adaptive radiation that took place over a period of 100 million years the dinosaurs had become the dominant vertebrates on Earth. They were an extremely diverse group in body shape, size and habitat, and included the largest animals ever to inhabit land. The Earth's climate was warm and constant, much like the present conditions in the tropics. The Age of the Dinosaurs came to an end when a cataclysmic event led to the mass extinction of life.

A number of different theories have been proposed to explain the extinction of the dinosaurs:

- ✓ There is good evidence that there was a marked change in climate and land levels and there may have also been a marked drop in temperature over much of the Earth. The dinosaurs may have been particularly sensitive to such climatic changes partly because of the large amounts of food they needed to stay alive.
- ✓ Another theory suggests that an asteroid, 10 kilometres in diameter, crashed onto the Earth 65 million years ago. The impact could have triggered volcanic eruptions leading to the release of vast quantities of dust that might have blocked out the sun for a few years, preventing plants from being able to photosynthesise and leading to mass starvation, particularly among larger animals. It is also less well known that almost half the species of marine non-vertebrates also became extinct during this period.

Principles of Taxonomy

It is believed that there are between 3 to 30 million species of living organisms on Earth but, to date, only about two million different kinds of organisms have been described and identified. The sorting of living organisms into groups of a manageable size is known as taxonomy or classification. You should know about the principles of modern classification that show how organisms may be related through evolution by the number of common characteristics they share.

Organisms are grouped in a hierarchical scheme

Taxonomy is the scientific study of the diversity of living organisms.
A classification system places order and a general plan upon the immense diversity of life.

When describing living organisms taxonomists look for differences and similarities between them.

- ✓ If organisms have evolved along the same lines similarities may occur. For example, the different mouthparts of insects. Some mouthparts are used for piercing e.g. female mosquito; others for chewing vegetation e.g. locust; yet others for sucking liquid food e.g. butterfly. These different mouthparts all follow a similar basic pattern. The study of the physical appearance of organisms allows biologists to group organisms together according to the extent to which they look alike.
- ✓ It is also possible to study the extent to which organisms are related to one another. Organisms with common ancestors are classified close together and those that are more distant relatives are classified further apart. Here, biochemical features of protein and DNA composition are more important than physical resemblance.

In a system of classification, scientists place similar organisms closely together and dissimilar ones further apart. A classification system based on large groups being divided up into progressively smaller groups is said to be hierarchical.
The natural classification used today was devised by the Swedish scientist, Linnaeus, in the 18thCentury. In this scheme organisms are grouped together according to their basic similarities. A hierarchical system has been devised to distinguish large groups of organisms with a series of rank names to identify the different levels within the hierarchy. A **taxon** is a level in this classification hierarchy and is a collection of organisms sharing some basic features.

Starting with the smallest group:

- **species** - a group of organisms which share a large number of common characteristics and which can interbreed to produce fertile offspring. (Fertile offspring are only produced when homologous chromosomes can pair at meiosis.) Examples are *Locusta* (Locust), *Rosa* (Rose).
- **genus** – a group of species that are very closely related. *Locusta migratoria* (migratory locust), *Rosa canina* (Dog rose).
- **family** - a grouping of similar genera (plural of genus), e.g. Rosaceae (Roses)
- **order** – grouping of related families, e.g. Orthoptera (includes locusts, grasshoppers and crickets).
- **class** – a grouping of similar orders, e.g. Insecta (Insects).
- **phylum** – a large grouping of all the classes that share some common features, e.g. Arthropods (includes insects, spiders, centipedes and millipedes, crustaceans).
- **kingdom** – the largest taxonomic grouping, e.g. animals, plants.

- ✓ Moving up the hierarchy from species to kingdom means that the relationship between organisms in the groups decreases.
- ✓ Moving down the hierarchy from kingdom to species means that the organisms are more closely related.

Binomial system

Many living organisms have common names, which may differ from one part of a country to another. This can be confusing and even more confusing if a particular species needs to be named and described in a scientific research paper, which may be read by a scientist from a different country.
To overcome the problem, organisms are named according to the **binomial system.** This system was introduced by Linnaeus in 1753 and is based on using Latin as an international language. Each organism is given two names, the name of its genus and the name of its species. This system means that an organism is given precise identification world-wide, whereas the common name is not. The binomial system is still successfully used today because not only does each particular organism have its own unique scientific name but it also allows biologists to recognise that that two species are closely related, e.g. *Panthera leo* (lion) and *Panthera tigris* (tiger).

When using the binomial system certain rules must be followed:
- ✓ The genus name is the first word and always has a capital letter.
- ✓ The species name always comes second and starts with a small letter.
- ✓ The first time the scientific name is used in a text it should be written out in full e.g. *Panthera tigris*
- ✓ The genus name can then be abbreviated, *P.tigris.*
- ✓ Both names should be written in italics.

The five kingdom classification

Until recently it was common to divide all living organisms into two kingdoms, plants and animals. Because there are some organisms which can be classified in both kingdoms and some which do not fit into either, the five-kingdom classification was proposed.

Prokaryote kingdom   Protoctist kingdom   Animal kingdom   Fungus kingdom   Plant kingdom

Prokaryotae

The prokaryotae are unicellular organisms and include bacteria and blue-green algae (Cyanobacteria). They have no internal cell membranes; no nuclear membrane; no endoplasmic reticulum, no mitochondria and no Golgi body. They possess a cell wall but it is not made of cellulose.

Protoctista

The members of this kingdom are mostly small eukaryotic organisms, with membrane-bound organelles and a nucleus with a nuclear membrane. In this kingdom are found the organisms that are neither plants nor animals nor fungi. The kingdom includes algae, the water moulds, the slime moulds and the protozoa.

Fungi

The members of this kingdom are eukaryotic, the body consisting of a network of threads called hyphae, forming a mycelium. There is a rigid cell wall made of chitin. They do not have photosynthetic pigments and feeding is heterotrophic; all members of the group are either saprophytic or parasitic. In some sub groups, the hyphae have no cross-walls, but in others cross-walls, or septa, are present. Reproduction is by spores that lack flagella. Examples are *Penicillium,* yeast, mushroom.

Plantae

The members of this kingdom are multicellular and carry out photosynthesis. The cells are eukaryotic, have cellulose walls, vacuoles containing cell sap, and chloroplasts containing photosynthetic pigments. The main plant phyla include:

- ✓ non- flowering plants
- ✓ mosses and liverworts
- ✓ ferns
- ✓ conifers
- ✓ flowering plants

The flowering plants (Angiosperms) are the most dominant plant group on Earth. They include all our major crops and are therefore an important food source. Their flowers have seeds that are enclosed in a fruit formed from the ovary wall.

Animalia

Members of the animal kingdom are multicellular, heterotrophic, eukaryotes, with cells lacking a cell wall, and show nervous coordination.

- ✓ Each phylum contains organisms grouped together according to a basic blueprint.
- ✓ Some phyla contain many more species than others.
- ✓ The animal kingdom is divided into two main groups:
- ✓ Non-chordates, often called invertebrates
- ✓ Chordates - all but the simplest of the chordates have a vertebral column and are therefore referred to as vertebrates.
- ✓ 95% of all animals are invertebrates and only 5% are vertebrates.

The animal kingdom is divided into 20 major phyla and several minor ones. They include:

- ✓ invertebrates
- ✓ jellyfish
- ✓ flatworms
- ✓ roundworms
- ✓ segmented worms
- ✓ molluscs
- ✓ arthropods
- ✓ starfish
- ✓ vertebrates (fish, amphibians, reptiles, birds and mammals)

At this level is not necessary to study even the main features of all these phyla. However, we will study in outline only the following three animal phyla, annelids, arthropods and vertebrates, *as they are encountered in other sections of this unit. i.e. gaseous exchange, circulation system, reproduction.*

Selected animal phyla

Annelids

There are 8,000 named species of annelids. They include earthworms, leeches, lugworms.
All members of the phylum have the following common features:

- ✓ a long, thin segmented body, the segments being visible externally as rings
- ✓ a body divided internally by partitions (septa)
- ✓ a fluid filled body cavity
- ✓ a hydrostatic skeleton
- ✓ a head end with a primitive brain and a nervous system running the length of the body
- ✓ specialised segments responsible for different functions, e.g. reproduction, excretion
- ✓ a thin permeable skin, through which gaseous exchange occurs
- ✓ a closed circulatory system containing an oxygen carrying pigment.

Arthropods

The arthropods are the most numerous and most successful of all the animals phyla. They include millipedes and centipedes, crustaceans, spiders and insects. The latter are by far the most successful class within the phylum with over one million species.

The arthropods have the following common features:

- ✓ a body divided into segments
- ✓ a body further divided into head, thorax and abdomen
- ✓ a well developed brain
- ✓ a hard outer exoskeleton made of chitin
- ✓ paired jointed legs
- ✓ an open circulatory system
- ✓ a cavity which surrounds the body organs.

Two important evolutionary developments are:

- **jointed legs** modified to perform a variety of functions, including walking, swimming, jumping, feeding, reproduction and where present, ventilation of the gills.
- **exoskeleton** – the outermost layer of cells of the body secretes a thick cuticle, which consists mainly of chitin. This performs several functions:
- ✓ protection of internal organs
- ✓ protection from predators
- ✓ provides a point of attachment for muscles
- ✓ support – for small animals a hollow tubular structure surrounding the body provides greater support than a solid cylindrical rod within it (an endoskeleton as in vertebrates) made from the same quantity of material
- ✓ in most terrestrial arthropods the exoskeleton is covered with a layer of wax which reduces water loss.

The one main disadvantage of the exoskeleton is that it is fixed in size and does not grow with the animal. This contrasts with the vertebrate endoskeleton which increases in size as the body grows. In order to grow an arthropod must periodically shed its exoskeleton (ecdysis). This leaves the animal especially vulnerable as the new exoskeleton hardens.

All arthropods have the same basic features listed above but are subdivided into four classes according to the differences between the animals in each classes. The main difference involves the number of legs that the animals possess:

- Millipedes and centipedes – they have many pairs of legs, one or two per segment (centipedes have one pair of legs per segment, millipedes have two pairs of legs per segment).
- Crustacea - have between 10 to 20 pairs of legs.
- Spiders - have four pairs of legs.
- Insects - have three pairs of legs.

The insect class

It is an incredible fact that 75% of all animals are insects! They are by far the most successful group of animals on Earth. In many ways they are more successful than humans.
Insects occupy most habitats and have also conquered the air.

Insects belong to the arthropod phylum but are subdivided as insects because they have the following specific features not found in other arthropods:

✓ They have three pairs of legs, one pair to each of the segments of the thorax.
✓ The head has a pair of antennae and compound eyes.
✓ Gas exchange is by gills in aquatic insects and tracheae in terrestrial forms.
✓ Many insect species have evolved wings, making them the only invertebrates that can fly. They are powerful flying machines and it is this power of flight that has played a large part in their success.

The presence in the adult of two pairs of wings and six legs is a diagnostic feature of insects. In the evolution of some insect groups these features may have been secondarily lost. This has happened in the evolution of fleas and lice.

Vertebrates

There are 60,000 named species of Chordates. Examples include frogs, snakes, eagles and humans.

Vertebrates possess:
✓ a vertebral column or backbone.
✓ a well-developed brain, enclosed in a cranium.

The vertebrates are subdivided into five classes:
✓ fish – aquatic forms with scales, fins and gills.
✓ amphibians – these were the first land vertebrates, partly terrestrial and partly aquatic. They have a soft, moist skin. The eggs are fertilised externally in water where they also develop. Young (larvae) are aquatic and have gills, adults are usually terrestrial and have simple lungs.
✓ reptiles – mainly terrestrial and have a dry skin with scales. They have lungs. The eggs are fertilised internally, covered with a shell and laid on land.
✓ birds – similar to reptiles in many ways. Differences are mainly due to the ability to fly and the development of feathers, with fore-legs developed as wings. They have lungs. Eggs have hard shells.
✓ mammals – skin with hair. Young are born alive and are fed on milk. Have lungs. They are further subdivided into two groups:

- marsupials e.g. kangaroo – young are born at a very immature state and develop in female's pouch.
- placentals – young undergo considerable development in the mother's womb, and receive nourishment via the placenta before they are born.

Evidence of common ancestry

To summarise, taxonomy is the study of grouping or classifying organisms and is concerned with:
✓ discovering and describing biological diversity
✓ investigating evolutionary relationships between organisms
✓ classifying these organisms so as to reflect these relationships.

The theory of evolution suggests that widely separated groups of organisms share a common ancestor. Therefore it would be expected that they share certain basic structural features. How similar they are should indicate how closely related they are in terms of evolution. Groups with little in common are assumed to have diverged from a common ancestor much earlier in geological history than groups which have a lot in common.

Pentadactyl limb

In deciding how closely related two organisms are a biologist needs to look for structures that, though they may serve quite different functions, are similar in structure, suggesting a common origin. Such structures are said to be **homologous**. A good example of this is the **pentadactyl limb** of vertebrates. A pentadactyl limb has five digits (or fingers). It is found in the four classes of terrestrial vertebrates, amphibians, reptiles, birds and mammals. The structure of the limb is basically the same in all the classes. However, the limbs of the different vertebrates have become adapted for different functions, such as grasping, walking, swimming and flying, in a selection of vertebrates.
Examples of the pentadactyl limb modified for different functions are the human arm, the wing of a bat, the flipper of a whale, the wing of a bird, the leg of a horse.
Using information such as this it is possible to construct evolutionary trees where the end products of evolution have certain structural features in common with each other and with the ancestral stock from which they arose. The more similar two organisms are the more recently they are assumed to have diverged.
However, there is a possible danger to assume that just because two animals look similar that they are related. Consider a shark, a porpoise and a penguin. One is a fish, one is a bird and the other is a mammal. By studying the skeleton of the fore-limbs it is possible to deduce that the dolphin and the penguin have modified pentadactyl limbs, but the fish does not. The shark and dolphin are animals that have similar fore-limbs because they live in similar environments and have become adapted to that environment, not because they have a common ancestor. That is, the structures, limbs, are performing the same function. Such structures are described as **analogous**. Another example is the wings of birds and insects.

Biochemical techniques

In recent years protein and DNA analysis has been used to confirm evolutionary relationships.
✓ The sequence of amino acids has been analysed in part of the fibrinogen molecule of various mammals. The sequence has been found to differ in varying degrees from one species to another and this has enabled scientists to draw up a possible evolutionary tree for mammals.
✓ On occasions it may be sufficient to use techniques, such as chromatography and electrophoresis, to separate and compare particular chemical substances present in different species. This may reveal similarities or differences which indicate how closely related the species are.
✓ It is possible to compare the sequence of bases in different organisms. The more alike the sequences, the closer the organisms are presumed to be in evolution. This technique is called DNA hybridisation.

- ✓ This comparison of DNA from a variety of living species can give a quantitative indication of how closely the genes are related. To work out how closely related two species of primates are, such as humans and chimpanzees, the DNA strands from both species are extracted, separated and cut into fragments. The fragments from the two species are then mixed and analysed. This technique gives results which show that chimpanzees and humans have 97.6% DNA in common, whereas humans and rhesus monkeys have 91.1% of DNA in common. Recent studies using this technique have also shown that the hippopotamus and whale are closely related.

2.2 Adaptations for gas exchange

All living organisms exchange gases with the environment. They need oxygen to convert organic molecules, such as glucose, into energy by the process of respiration. In turn, waste gases have to be removed. Organisms live in different environments; some live in water and others on land. An aquatic environment is fairly constant but life on land can be more extreme, ranging from the rarified atmosphere at the top of a mountain, to the intense heat of the arid desert. In order to survive, living organisms have adapted in different ways.

Overcoming problems associated with increase in size

Gas exchange is the process by which oxygen reaches cells and carbon dioxide is removed from them. It should not be confused with respiration.

Animals and plants have evolved special gas exchange surfaces so that diffusion of gases into and out of cells can take place rapidly and efficiently. Gas exchange surfaces such as the gills of a fish, the alveoli in the lungs of a mammal, the trachea of an insect and the spongy mesophyll cells in the leaves of a plant, are all excellent gas exchange surfaces.

In order to achieve the maximum rate of diffusion a respiratory surface must:

✓ have a sufficiently **large surface area** relative to the volume of the organism to satisfy the needs of the organism
✓ be **thin**, so that diffusion paths are short
✓ be **permeable** to allow the respiratory gases through
✓ **moist** to allow a medium in which the gases dissolve before diffusion
✓ maintain a concentration gradient.

- In simple **single-celled** organisms such as *Amoeba*, the cell membrane is the gas exchange surface. The organism lives in water and the diffusion of gases occurs over the whole of the body surface. A single cell has a large surface area compared to its volume. It is said to have a large surface area to volume ratio. The efficiency of gaseous diffusion satisfies the needs of the organism. The membrane is thin, moist, and the diffusion paths are short.

- There must be a limit to cell size and a point is reached where the diffusion path is so long that the process of diffusion becomes inefficient. In terms of evolution, the only way that organisms could become larger was to aggregate cells together, that is, to become **multicellular**. However, the larger the organism the smaller the surface area to volume ratio. Also, materials need to be exchanged between different organs as well as between the organs and the environment. This means that the gaseous requirements, provided by diffusion through the cell surface is insufficient to meet the needs of the organism. In effect, the process of diffusion is too slow.

- **Simple multicellular animals,** such as worms, have modest oxygen requirements because they are slow moving and so have a very low metabolic rate. Oxygen and carbon dioxide diffuse across the skin surface and they do not have any special gas exchange organs.

✓ **Flatworms** are aquatic animals that have evolved a flattened shape. This considerably increases the surface area to volume ratio and ensures that no part of the body is far from the surface, i.e. diffusion paths are short.
✓ **Earthworms** are terrestrial organisms that have developed a tubular shape and are restricted to the damp environment of soil. Its elongated shape provides it with a large surface area to volume ratio compared to a compact organism of the same volume. Although it does not require a special surface for gaseous exchange it does need to keep its skin moist by secreting mucus onto the surface.

However, once oxygen is inside the body it needs to be transported some distance to the many internal cells. The earthworm has a closed blood system containing blood within vessels. The blood also contains a respiratory pigment for the transport of oxygen.
Oxygen diffuses into the blood capillaries beneath the skin surface and is carried in vessels to the cells with carbon dioxide being transported in the opposite direction. In this way the blood system maintains a diffusion gradient at the respiratory surface.

- **Larger and more advanced multicellular animals,** such as insects, fish, reptiles and mammals, have a high metabolic rate i.e. they need more energy and have a high requirement for oxygen. Also, the larger the organism the smaller the surface area to volume ratio. With an increase in size and specialisation, tissues and organs became more dependent on one another. In order to make gas exchange more efficient these organisms have developed **a specialised exchange surface** to compensate for the increased oxygen demand.

✓ Terrestrial **insects** have air-filled tubes called tracheae.
✓ In aquatic insects and **fish** the exchange surfaces for respiratory gases take the form of **gills**.
✓ Terrestrial animal groups such as **birds, reptiles and mammals** have developed **lungs.**

All of these different mechanisms need a means of **ventilation** to supply the respiratory surfaces with a fresh supply of oxygen and to maintain diffusion gradients. That is, the function of a ventilation mechanism is to move the respiratory medium, air or water, over the respiratory surface.

These different animal groups have also developed:
✓ **an internal transport system** – provided by a blood circulation system to move gases between respiring cells and the respiratory surface.
✓ **a respiratory pigment** in the blood – to increase its oxygen carrying capacity.

Gas exchange in the fish

Aquatic organisms have a problem with gaseous exchange because water contains far less oxygen than air and the rate of diffusion in water is slower. Also, water is a more dense medium than air and does not flow as freely. As fish are very active they need a good supply of oxygen. In fish gaseous exchange takes place across a special surface, the **gill,** over which a one-way current of water is kept flowing by a specialised pumping mechanism. The density of the water prevents the gills from collapsing and lying on top of each other, which would reduce the surface area. Gills are made up of many folds, providing a large surface area over which water can flow, and gases can be exchanged.

Fish are divided into two main groups according to the material that makes up their skeleton.

- **Cartilaginous fish** e.g. sharks, have a skeleton made entirely of cartilage. Nearly all live in the sea. Just behind the head on each side are five gill clefts which open at gill slits. Water is taken into the mouth and is forced through the gill slits when the floor of the mouth is raised. Blood travels through the gill capillaries in the same direction as the sea-water. Gas exchange in such a **parallel flow** is relatively inefficient.
- **Bony fish** have an internal skeleton made of bone and the gills are covered with a flap called the operculum. Bony fish inhabit both fresh and sea-water and are by far the most numerous of aquatic vertebrates. There are about seven times more species of bony fish than cartilaginous species. Gas exchange involves a **counter current flow** arrangement whereby blood in the gill capillaries flows in the opposite direction to the water flowing over the gill surface (see next page).

In bony fish there are four pairs of gills in the pharynx (throat) and each gill is supported by a gill arch. Along each gill arch are many thin plates called gill lamellae and on these are the gas exchange surfaces, the gill plates. Out of water the gill collapses as the gill lamellae lie on top of each and stick together. However, in water they are supported and provide a large surface area. The gill plates contain blood capillaries and the oxygen passes through the gill plates into the capillaries and carbon dioxide passes out into the water.

❑ ***Action*** 1. Draw a diagram of a horizontal section through the pharynx and gill region
2. Draw a labelled diagram of the detailed structure of the gills of a bony fish.

To summarise: Gills provide
- a specialised area rather than using the whole body surface
- a large surface extended by the gill filaments
- an extensive network of blood capillaries to allow efficient diffusion and haemoglobin for oxygen carriage.

To increase efficiency water needs to be forced over the gill filaments by pressure differences so maintaining a continuous, unidirectional flow of water. A lower pressure is maintained in the opercular cavity than in the bucco-pharynx. The operculum acts as both a valve, permitting water out, and as a pump drawing water past the gill filaments. The mouth also acts as a pump.
The **ventilation mechanism** for forcing water over the gill filaments operates as follows:-

mouth	opens
operculum	closes
floor of buccal cavity	lowered
volume	increases
pressure	decreases

water flows in

Counter current flow.
The orientation of the gas-exchange surfaces is such that as the water passes from the pharynx into the opercular chamber, it flows between the gill plates in the opposite direction to the blood flow. This increases efficiency because the diffusion gradient between the adjacent flows is maintained over the whole length of the gill filament. That is, the blood always meets water with a relatively higher oxygen content. This system allows the gills of a bony fish to remove 80% of the oxygen from water. This is three times the rate of extraction of oxygen from air in human lungs. This high level of extraction is essential to fish as there is around 25 times less oxygen in water, compared to air.

How vertebrate groups have adapted to gaseous exchange on land

Vertebrates include the five classes, fish, amphibia, reptiles, birds and mammals. Life is thought to have evolved in water with animals adapting in order to colonise the land and some becoming highly modified for flight. Gills do not function out of water and there was a need for animals to evolve a different form of gaseous exchange surface, the lung. Birds and mammals are particularly active and adapted for exchange with air, a less dense medium, instead of water, so have internal lungs to minimise loss of water and heat. The ventilation and structure of the human lung is dealt with in detail.

Amphibians

The amphibians include frogs, toads and newts. This group was probably the first vertebrate group to colonise the land. Frogs are typical of amphibians in that they live in moist habitats as they require water for fertilisation (fertilisation is external and gametes swim in water – see section 2.4). The larvae (tadpoles) also live in water and have gills. The transition from larva to land living adult involves great changes in the body form, known as metamorphosis. The inactive adult uses the moist skin as a respiratory surface and this provides sufficient oxygen for its needs. However, when active, as in mating, the frog uses lungs as a respiratory surface.

Reptiles

Reptiles include crocodiles, lizards, and snakes. The few present-day reptiles are descendants of a once very successful group of animals, including the dinosaurs, which dominated the Earth about 200 million years ago. They are far better suited to life on land than amphibians. Reptiles can move on all four limbs without the trunk of the body touching the ground. Pairs of ribs project from the vertebrae (backbone). Ribs provide support and protection to the organs in the body cavity. Ribs are also involved in the ventilation of the lungs. The lung also has a more complex internal structure than that of amphibians with the in-growth of tissues increasing the surface area for gas exchange.

Birds

The lungs of birds have an internal structure similar to that of mammals. However, large volumes of oxygen are needed to provide the energy for flight. Ventilation of the lungs in birds is far more efficient than in other vertebrates. It is assisted by a system of air sacs connected to the lungs. The air sacs function as bellows. When a bird breathes in, any air that remains in the lungs from a previous ventilation, is sucked into the sacs. The lungs are filled with fresh air, avoiding the dead space that occurs in the lungs of other vertebrates, such as mammals. Ventilation of the lungs is brought about by the movement of the ribs, there is no diaphragm as in humans. During flight the action of the flight muscles ventilates the lungs. Gaseous exchange is very efficient with virtually no residual gas remaining in the lung tubes.

Insects have evolved a different mechanism of gaseous exchange

Most insects possess wings and are extremely efficient in the air. Flight involves a high expenditure of energy and so insects need a good supply of oxygen. They have evolved a different system of gaseous exchange to other land animals. Gas exchange occurs through paired holes, called spiracles, running along the side of the body. The spiracles lead into a system of branched chitin lined air-tubes called tracheae. The spiracles can open and close like valves. This allows gaseous exchange to take place and also reduces water loss.

Resting insects rely on diffusion to take in oxygen and to remove carbon dioxide. During periods of activity, such as during flight, movements of the abdomen ventilate the tracheae. The end of the tracheael branches are called tracheoles; here gaseous exchange takes place resulting in oxygen passing directly into the cells.

- ❑ **Action** Draw a simple diagram to illustrate a branched trachea.

In order to survive on land terrestrial organisms need to conserve water. But at the same time the gaseous exchange surfaces need to be moist for efficient diffusion to take place. Therefore organs of gaseous exchange need to be retained in the body. When you review this section you should consider the advantages of the internal lung compared to the external gill as an adaptation to life on land.

Human respiratory system

The structure of the human respiratory system

The lungs are enclosed within an airtight compartment, the **thorax,** at the base of which is a dome-shaped sheet of muscle called the **diaphragm.** Air is drawn into the lungs via the trachea. The lungs consist of a branching network of tubes called **bronchioles** arising from a pair of **bronchi.**

❑ ***Action*** Draw a labelled diagram of the respiratory system.
(include the following labelled parts: epiglottis, trachea, bronchi, bronchioles, alveoli, pleural membranes, ribs, intercostal muscles, diaphragm.)

Gas exchange in the alveolus

The gas exchange surfaces are the alveoli (air sacs) which provide a very large surface area relative to the volume of the body. They are well suited as a gas exchange surface because:

✓ Their surfaces are moist for gases to dissolve
✓ The walls are thin helping diffusion by providing a short diffusion path
✓ Each alveolus is covered by an extensive capillary network to maintain diffusion gradients, because blood is always taking oxygen away from the alveolus and returning with carbon dioxide.

❑ ***Action*** Draw a labelled diagram of a single alveolus

Table showing the % composition of air in the lungs.

Gas	Inspired air	Alveolar air	Expired air
Oxygen	20.95	13.80	16.40
Carbon dioxide	0.04	5.50	4.00
Nitrogen	79.01	80.70	79.60
Water	variable	saturated	saturated

The percentage of oxygen in the alveolus is less than in inspired air because the latter mixes with air already in the lungs and this has a lower percentage of oxygen.

Ventilation of the lungs

Mammals ventilate their lungs by negative pressure breathing, forcing air down into the lungs. That is, if air is to enter the lungs then the pressure inside them must be lower than atmospheric pressure.

	Inspiration (breathing in)	Expiration (breathing out)
external intercostal muscle	contracts	relaxes
ribs	up and out	down and in
diaphragm	contracts and flattens	relaxes
volume of thorax	increases	decreases
pressure in thorax	decreases	increases
outside air (atmospheric) pressure	Greater therefore air moves in	Less therefore air moves out

Surrounding each lung and lining the thorax are pleural membranes between which is a cavity containing pleural fluid. When breathing this fluid acts as a lubricant allowing friction-free movement against the inner wall of the thorax. To prevent the alveoli from collapsing when breathing out an anti-sticking chemical called a surfactant covers their surfaces and reduces the surface tension.

Gas exchange in plants

In the same way that animals need to respire all the time so do plants! However, plant cells containing chloroplasts are also capable of carrying out the process of photosynthesis. During the day plants both respire and photosynthesise. Most of the carbon dioxide they need for photosynthesis diffuses into the leaves from the atmosphere. However, some of the carbon dioxide is provided by the respiration. Most of the oxygen produced by photosynthesis diffuses out of leaves. During the night plants respire only and need a supply of oxygen from the atmosphere. Although some oxygen enters the root by diffusion most gas exchange takes place in the leaves.

The leaf as an organ of gaseous exchange

The structure of the leaves of flowering plants is related to their function of gaseous exchange.
To enable gaseous exchange to take place efficiently:

- the leaf blade is thin and flat with a large surface area
- the spongy mesophyll tissue allows for the circulation of gases
- the plant tissues are permeated by air spaces
- the stomatal pores permit gas exchange.

❑ ***Action*** Draw a labelled diagram of T.S. angiosperm leaf.
(include the following labels: cuticle, epidermis, palisade mesophyll, spongy mesophyll, vascular bundle, air space, stomata, guard cells.)

Gases diffuse through the stomata along a concentration gradient. Once inside the leaf the gases in the sub-stomatal air chambers diffuse through the intercellular spaces between the mesophyll cells and into the cells. The direction of diffusion depends on the environmental conditions and the requirements of the plants. It is the net exchange of carbon dioxide and oxygen in relation to respiration and photosynthesis that matters.

Adaptations of the leaf for photosynthesis

To ensure the efficient absorption of light the leaf shows the following adaptations:

- Leaves have a large surface area to capture as much sunlight as possible.
- Leaves can orientate themselves so that they are held at an angle perpendicular to the sun during the day to expose the maximum area to the light.
- Leaves are thin to allow light to penetrate the lower layers of cells.
- The cuticle and epidermis are transparent to allow light to penetrate to the mesophyll.
- Palisade cells are elongated and densely arranged in a layer, or layers.
- The palisade cells are packed with chloroplasts and arranged with their long axes perpendicular to the surface.
- The chloroplasts can rotate and move within the mesophyll cells. This allows them to arrange themselves into the best positions for the efficient absorption of light.
- The intercellular air spaces in the spongy mesophyll allow carbon dioxide to diffuse to the cells and oxygen can diffuse away.

Stomata

Stomata are small pores found on the lower surface of a leaf. Each pore is bounded by two guard cells. Guard cells are unusual in having chloroplasts and unevenly thickened walls, with the inner wall being thick and the outer wall thin. The stomata allow exchange of gases between the atmosphere and the internal tissues of the leaf. However, water also evaporates from a plant though the stomata. Plants wilt if they lose too much water. Since light hits the upper surface of the leaf, confining stomata to the lower surface reduces water loss. The presence of a waxy cuticle on the upper surface also reduces water loss significantly. In most plants the stomata close at night.

- ❑ **Action** Draw stomata in T.S and in surface view.

The opening and closing mechanism of stomata

- Stomata are pores in the epidermis, each bordered by two guard cells. The guard cells differ from the other epidermal cells in that they possess chloroplasts. The inner wall of each guard cell is thicker than the outer wall.
- Guard cells around the stomata can change shape to open and close the stomata so helping to control gas exchange and water loss.
- An inevitable result of the necessity for the inside of the leaf to be open to the atmosphere for the exchange of gases is that water is lost from the leaves by the process of transpiration. In general, stomata are open during the day and closed at night. This prevents the plant from needlessly losing water when the light intensity is insufficient for photosynthesis to take place.

Guard cells change shape because of changes in turgor.
If water enters the guard cells they become turgid and the pore opens
If water leaves the guard cells they become flaccid and the pore closes

- ❑ ***Action*** Draw labelled diagrams of stomata in the open and closed conditions.

Several theories have been proposed to explain how this happens. The following is one theory.

During the day the mechanism for stomatal opening occurs as follows:

- potassium ion (K^+) pumps in the cell membranes of the surrounding epidermal cells actively transport K^+ ions into the guard cells
- stored starch is converted to malate
- the water potential in the guard cells is lowered (becomes more negative) and water enters by osmosis
- the guard cells become turgid and curve apart because their outer walls are thinner than the inner walls, so the pore widens.

The reverse process occurs at night and the pore closes.

- ❑ ***Action*** Describe the mechanism for stomatal closure.

Xerophytes

Xerophytes are plants that have adapted to living under conditions of low water availability. They have modified structures to prevent excessive water loss. Xerophytes may close stomata during the day and open them at night in order to conserve water. During periods of water stress (drought) the stomata of mesophytes may also close during the day.

2.3 Adaptations for transport

Oxygen enters the blood by diffusion through the respiratory surfaces, such as the alveoli of the lungs. However, movement of oxygen around the body by diffusion alone would be a very slow process. Even an inactive earthworm could only obtain about 10% of its oxygen requirement if it had to rely on diffusion to distribute the gas throughout its body. Almost all multicellular organisms need a rapid means of transport provided by an internal transport system. They need a transport system to carry oxygen, nutrients, carbon dioxide and waste products to and from the exchange surfaces.

Mammals, such as humans, have a circulatory system consisting of:

- ✓ blood which is transported through the system
- ✓ blood vessels which are the tubes that carry the blood
- ✓ a heart to pump the blood through the blood vessels.
- ✓ Typically the blood contains a respiratory pigment (absent in insects) which increases the volume of oxygen that can be transported.

An open or closed system

Insects have an open blood system.

In an open blood system blood is pumped at relatively low pressure from one main long, dorsal (top) tube shaped heart running the length of the body. The blood is pumped out of this heart into spaces, collectively called a haemocoel, within the body cavity. The blood bathes the tissues directly, exchange of materials takes place, and there is little control over the direction of circulation. Blood slowly returns to the heart. Here valves and waves of contraction of the muscle wall move the blood forward to the head region where the open circulation is started again. There is no respiratory pigment in the insect as the blood of an insect does not transport oxygen. Oxygen is transported directly to the tissues via the tracheae.

In a closed circulation system the blood circulates in a continuous system of tubes, the blood vessels. Blood is pumped by a muscular heart at a high pressure resulting in a rapid flow rate. Organs are not in direct contact with the blood but are bathed by tissue fluid seeping out from thin-walled capillaries. The blood contains a blood pigment which carries oxygen.

The earthworm is an example of an animal with a closed circulation system. It has dorsal and ventral vessels running the length of the body and these are connected by five 'pseudohearts'. Blood moves through the vessels by the pumping action of the 'pseudohearts'.

Single and double circulations

Closed circulation systems are of two types, depending on whether the blood passes through the heart once or twice in each circulation of the body.

- Fish have a single circulation. The heart pumps deoxygenated blood to the gills, oxygenated blood is then carried to the tissues, from there deoxygenated blood returns to the heart. Blood goes once through the heart during each circuit of the body.
- Mammals have a double circulation.
- ✓ The pulmonary circulation - the right side of the heart pumps deoxygenated blood to the lungs. Oxygenated blood then returns to the left side of the heart.
- ✓ Systemic circulation – the left side of the heart pumps the oxygenated blood to the tissues. Deoxygenated blood then returns to the right side of the heart.

  In each circuit the blood passes through the heart twice, once though the right side and once through the left side.

  - ❑ **Action** Draw labelled diagrams to show the circulation of blood in a fish and a human. Name the main blood vessels associated with the human heart – vena cava, pulmonary artery, pulmonary vein, dorsal aorta.

Human circulatory system and associated blood vessels

The human circulatory consists of a closed double circulation and a heart with two atria and two ventricles. The transport system incorporates a **pump (**the heart) to sustain high pressure, **valves** to control the flow, and **vessels** to distribute the blood.
There are three main types of blood vessels; arteries, veins and capillaries.
Arteries carry blood away from the heart. The arteries have thick, muscular walls to withstand the high pressure of blood received from the heart. The contraction of the arterial muscles also helps to maintain pressure as the blood is transported further from the heart.
The arteries branch into smaller vessels called arterioles that further subdivide into thin-walled **capillaries**. The capillaries form a vast network which penetrates all the tissues and organs of the body. Blood from the capillaries collects into venules, which in turn empty blood into **veins**, from which it is returned to the heart.

❑ ***Action*** Draw and label a general plan of the circulatory system in the human.
(You are required to name only the main blood vessels associated with the heart.)

Structure and function of blood vessels

Arteries and **veins** have the same basic three-layered structure but the proportions of the different layers vary.
In both arteries and veins:

- the innermost layer is the **endothelium**, which is one cell thick and provides a smooth lining to reduce friction and provide a minimum resistance to the flow of the blood.
- the middle layer is made up of **elastic fibres** and **smooth muscle**. This layer is thicker in the arteries than in the veins to accommodate changes in blood flow and pressure as blood is pumped from the heart.
- the outer layer is made up of **collagen fibres** which are resistant to over-stretching.

Veins have larger diameters and thinner walls than arteries as the pressure and flow is reduced.
Veins have semi-lunar **valves** along their length to ensure flow in one direction (prevent back flow); these are not present in arteries apart from the aortic valves.

The **capillaries** are thin walled consisting only of a layer of endothelium so their walls are permeable to water and dissolved substances such as glucose. It is at the capillaries that the exchange of materials between the blood and the tissues takes place.
Capillaries have a small diameter and friction with the walls slows the blood flow. Although the diameter is small there are many capillaries in the capillary bed providing a large total cross-sectional area which further reduces blood flow. This low velocity in very thin walled vessels enhances their ability to exchange materials with the surrounding tissue fluid.
(It may at first seem that the blood should travel faster through capillaries than through arteries, because the diameters of the capillaries are much smaller. However, it is the total cross sectional area delivering the blood that determines the flow rate.)

❑ ***Action***
1. Draw labelled diagrams of an artery, vein and capillary.
2. Construct a detailed table comparing arteries and veins.

The heart

A pump to circulate blood is an essential feature of a circulatory system. The heart consists of a relatively thin walled collection chamber and a thick walled pumping chamber which are partitioned into two, allowing the complete separation of oxygenated and deoxygenated blood. The heart is, in effect, two pumps side by side.

The heart

The four-chambered heart is situated in the thorax between the two lungs. It consists largely of **cardiac** muscle, a specialised tissue that is capable of rhythmical contraction and relaxation of its own accord throughout a person's life. The heart muscle is said to be 'myogenic'.

❑ ***Action*** Draw a labelled diagram of the mammalian heart as seen in vertical section.

The cardiac cycle

The cardiac cycle describes the sequence of events in one heartbeat. The pumping action of the heart consists of alternating contractions (**systole**) and relaxations (**diastole**).
There are three stages to the cardiac cycle:

- Stage 1 – The right and left ventricles relax, the tricuspid and bicuspid valves open as the atria contract and blood flows into the ventricles.
- Stage 2 – The atria relax and the right and left ventricles contract together forcing blood out of the heart into the pulmonary artery and the aorta as the semilunar valves are opened. The tricuspid and bicuspid valves are closed by the rise in ventricular pressure. The pulmonary artery carries deoxygenated blood to the lungs and the aorta carries oxygenated blood to the various parts of the body.
- Stage 3 – The ventricles relax and the pressure in the ventricles falls. Blood under high pressure in the arteries causes the semilunar valves to shut, preventing blood from going back into the ventricles. Blood from the vena cavae and pulmonary veins enters the atria and the cycle starts again.

The following describes the flow of blood through the left side of the heart.

- The left atrium is relaxed and receives oxygenated blood from the pulmonary vein.
- When full the pressure forces open the bicuspid valve between the atrium and ventricle.
- Relaxation of the left ventricle draws blood from the left atrium.
- The left atrium contracts pushing the remaining blood into the right ventricle through the valve.
- With the left atrium relaxed and with the bicuspid valve closed the left ventricle contracts.
- The strong muscular walls exert a strong pressure and push blood away from the heart through the semilunar valves through the pulmonary arteries and the aorta.

❑ **Action** Describe the flow of blood through the right side of the heart.

- Both sides of the heart work together i.e. both ventricles contract at the same time, both atria contract together. One complete contraction and relaxation is called a heartbeat.
- After contraction, and the compartment has been emptied of blood, it relaxes, to be filled with blood once more.
- The ventricles contain more muscle than the atria and so generate more pressure to force the blood a greater distance.
- The left ventricle has a thicker muscular wall than the right ventricle as it has to pump the blood all round the body whereas the right ventricle has only to pump the blood a shorter distance to the lungs.

Pressure changes in the heart

- The highest pressure occurs in the aorta/arteries that show a rhythmic rise and fall corresponding to ventricular contraction.
- Friction with vessel walls causes a progressive drop in pressure. Arterioles have a large total surface area and a relatively narrow bore causing a substantial reduction from aortic pressure. Their pressure depends on whether they are dilated or contracted.
- The extensive capillary beds have a large cross sectional area. These beds create an even greater resistance to blood flow.
- There is a relationship between pressure and speed and the pressure drops further due to leakage from capillaries into tissues.
- The return flow to the heart is non-rhythmic and the pressure in the veins is low but can be increased by the massaging effect of muscles.

❑ ***Action*** 1. Draw labelled diagrams to illustrate the cardiac cycle.
2. Study graphs showing volume and pressure changes in the cardiac cycle.

Control of heartbeat

The heart muscle is **myogenic** i.e. the heartbeat is initiated from within the muscle itself and is not due to nervous stimulation.

- In the wall of the right atrium is a region of specialised cardiac fibres called the **sinoatrial node** (SAN) which acts as a **pacemaker**.
- A wave of electrical stimulation arises at this point and then spreads over the two atria causing them to contract more or less at the same time.
- The electrical stimulation is prevented from spreading to the ventricles by a thin layer of connective tissue. This acts as a layer of insulation (it is important that the muscles of the ventricles do not start to contract until the muscles of the atria have finished contracting).
- The stimulation reaches another specialised region of cardiac fibres, the **atrio-ventricular node** (AVN), which lies between the two atria and which passes on the excitation to specialised tissues in the ventricles.
- From the AVN the excitation passes down the **Bundle of His** to the apex. The Bundle branches into **Purkinje fibres** in the ventricle walls which carry the wave of excitation upwards through the ventricle muscle.
- The impulses cause the cardiac muscle in each ventricle to contract simultaneously from the apex upwards.

❑ ***Action*** Draw a labelled diagram of a vertical section through the heart to show the position of the sino-atrial node, atrio-ventricular node and the Bundle of His.

Blood and the transport of materials

In mammals a transport system provides the link between specialised areas for gas exchange and the cells which require oxygen and nutrients. As all cells are bathed in an aqueous medium, the delivery of materials to and from these cells is carried out largely in solution. The fluid in which the materials are dissolved or suspended is blood. Oxygen is carried as oxy-haemoglobin in the red blood cells. The release of oxygen involves the Bohr effect where the lowered pH due to dissolved carbon dioxide reduces the oxygen affinity of haemoglobin, causing it to release oxygen where it is most needed.

Composition of blood

Blood is a tissue made up of **cells** (45%) in a fluid **plasma** (55%).

- **plasma** is made up largely of 90% water, with soluble food molecules, waste products, hormones, plasma proteins, mineral ions and vitamins dissolved in it.
- **blood cells** which are of two main types:

✓ erythrocytes or red blood corpuscles
✓ leucocytes or white blood corpuscles.

Functions of blood

- Plasma - **transports** carbon dioxide, digested food products, hormones, plasma proteins, fibrinogen, antibodies etc. and also distributes heat.
- Red blood cells - these are filled with the pigment **haemoglobin**, are biconcave in shape and do not contain a nucleus. Their function is the **carriage of oxygen.**
- White blood cells - these are of two groups:

✓ granulocytes, which are **phagocytic**, have granular cytoplasm, lobed nuclei and **engulf** bacteria.
✓ agranulocytes, which produce **antibodies** and **antitoxins**, have clear cytoplasm and spherical nuclei.

The transport of oxygen

The partial pressure of oxygen (pO_2) is a measure of the oxygen concentration. The greater the concentration of dissolved oxygen the higher its partial pressure.

✓ Normal atmospheric pressure = 100 kiloPascals.
✓ Oxygen partial pressure = 21 kiloPascals.
(as oxygen makes up approximately 21% of the atmosphere).

Red blood cells load (pick up) oxygen in the lungs where the partial pressure is high and the haemoglobin becomes saturated with oxygen. The cells carry the oxygen as oxy-haemoglobin to the respiring tissues e.g. muscle, where the partial pressure is low (as oxygen is being used up in respiration to create energy). Oxy-haemoglobin then unloads its oxygen, that is, it **dissociates**.
A saturation graph of haemoglobin with increasing partial pressure of oxygen may be expected to show a straight line bisecting the two axes but, in fact samples of haemoglobin exposed to increasing partial pressures of oxygen show an **oxygen dissociation curve**.

❑ **Action** Draw an oxygen dissociation curve for the human adult.

An efficient respiratory pigment readily picks up oxygen at the respiratory surface and releases it on arrival at the tissues. Respiratory pigments have a high affinity for oxygen when the concentration is high but this is reduced when the concentration is low. The special properties of haemoglobin ensure that at high partial pressures of oxygen, as in the lungs, it combines readily with large amounts of the gas, i.e. haemoglobin becomes almost fully saturated with oxygen.
The amount of oxygen carried by haemoglobin depends not only on the partial pressure of oxygen but also the partial pressure of carbon dioxide.

At higher partial pressure of carbon dioxide the oxygen dissociation curve shifts to the right. This phenomenon is known as the **Bohr effect.** When oxygen reaches respiring tissues, such as muscle, the high partial pressure of carbon dioxide there, enables haemoglobin to unload its oxygen even more readily.
To summarise:

- When the respiratory pigment **haemoglobin** is exposed to a gradual increase in oxygen tension it absorbs oxygen rapidly at first but more slowly as the tension continues to rise. This relationship is known as the **oxygen dissociation curve**.
- The more the dissociation curve of haemoglobin is displaced to the right, the less readily it picks up oxygen, but the more easily it releases it.
- The more the dissociation curve of haemoglobin is displaced to the left, the more readily it picks up oxygen, but the less readily it releases.
- The release of oxygen from haemoglobin is facilitated by the presence of carbon dioxide, when the partial pressure of oxygen is high, as in the lung capillaries, oxygen combines with the haemoglobin to form oxyhaemoglobin.
- when the partial pressure of oxygen is low, as found in the respiring tissues, then the oxygen dissociates from the haemoglobin.
- when the partial pressure of carbon dioxide is high, haemoglobin is less efficient at taking up oxygen and more efficient at releasing it.

The dissociation curve of foetal haemoglobin

The blood of the foetus and the mother flow closely together in the placenta but rarely mix.
To enable the foetal haemoglobin to absorb oxygen from the maternal haemoglobin in the placenta the foetus has a haemoglobin that differs (in two of the four polypeptide chains) from the haemoglobin of the adult. This structural difference makes the foetal haemoglobin dissociation curve shift to the left of that of the adult. The foetal haemoglobin combines with oxygen more readily than does the mother's haemoglobin. That is, the foetal haemoglobin has a greater affinity for oxygen.

❑ ***Action*** Draw the oxygen dissociation curves for foetal haemoglobin.

Transport of oxygen in other animals

The chemical composition of haemoglobins is not the same in all animals. Some animals have become adapted to living in habitats where there are low levels of oxygen.

- The lugworm has a low metabolic rate and lives in the sand on the seashore (worm casts can be seen at low tide). The lugworm pumps seawater through its burrow, giving access to the limited amount of dissolved oxygen present. To enable it to load the oxygen more readily it has a haemoglobin with a dissociation curve very much to the left compared with a human haemoglobin dissociation curve.
- With increase in altitude there is a drop in atmospheric pressure. This is significant for animals, such as the llama, because the partial pressure of oxygen in the atmosphere is less at high altitude.

To compensate for this:

✓ The llama possesses haemoglobin which loads more readily with oxygen in the lungs. Haemoglobin of this sort has a dissociation curve to the left of normal haemoglobin.
✓ At high altitude the number of red cells in the blood of mammals increases.

Myoglobin

Myoglobin is far more stable than haemoglobin and its dissociation curve is far to the left of haemoglobin. At each partial pressure of oxygen, myoglobin has a higher percentage oxygen saturation than haemoglobin. Normally, the respiring muscle obtains its oxygen from haemoglobin. However, if the oxygen partial pressure becomes very low, as when exercising, the oxymyoglobin unloads its oxygen. Myoglobin therefore acts as an energy store in muscle.

Some carbon dioxide is transported in the red blood cells but most is converted in the red blood cells to bicarbonate which is then dissolved in the plasma. The chloride shift refers to the influx of chloride ions into the red blood cells to preserve electrical neutrality.

Transport of carbon dioxide - the chloride shift

Haemoglobin acts as a buffer helping to maintain the blood pH by removing hydrogen ions from solution. Carbon dioxide is transported in blood cells and plasma in 3 ways:

- 5% in solution in the plasma.(This is inadequate to meet the needs of most organisms.)
- 85% as hydrogencarbonate
- 10% in combination with haemoglobin to form carbamino-haemoglobin.

The following describes a series of reactions known as the **chloride shift**:

✓ Carbon dioxide diffuses into the red blood cell (RBC) and combines with water to form **carbonic acid**.
✓ Carbonic acid dissociates into H^+ and HCO^{3-} ions, the reaction being catalysed by **carbonic anhydrase**.
✓ HCO^{3-} ions diffuse out of the RBC into the plasma where they combine with Na^+ ions from the dissociation of sodium chloride to form sodium hydrogen carbonate.
✓ H^+ ions provides the conditions for the **oxyhaemoglobin** to dissociate into oxygen and haemoglobin.
✓ H^+ ions are buffered by their combination with haemoglobin and the formation of **haemoglobinic acid** (HHb).
✓ The oxygen diffuses out of the RBC into the tissues.
✓ To balance the outward movement of negatively charged ions, **chloride ions** diffuse in.
✓ This is known as the **chloride shift** and it is by this means that the **electrochemical neutrality** of the RBC is maintained.

❑ ***Action*** Using a diagram or flowchart summarise the main chemical events that take place in a red blood cell on reaching the tissues.

Intercellular fluid

The **capillaries** are the site of exchange between the blood and the cells of the body. They are well adapted to allow the exchange of materials between the blood and the cells.

✓ they have thin, permeable walls
✓ they provide a large surface area for exchange of materials
✓ blood flows very slowly through the capillaries allowing time for exchange of materials.

❑ **Action** Draw a labelled diagram showing a capillary network.

Blood consists of the fluid plasma that carries the blood cells, dissolved materials and large molecules, called plasma proteins. The blood is contained in a closed system but fluid from the plasma escapes through the walls of the capillaries. This fluid is called **tissue fluid** and bathes the cells, supplying them with glucose, amino acids, fatty acids, salts and oxygen. The tissue fluid also removes waste materials from the cells.

Tissue fluid = plasma minus plasma proteins.

- The factors responsible for the movements of solutes and water into and out of the capillaries are blood pressure and diffusion.
- When blood reaches the arterial end of a capillary it is under pressure because of the pumping action of the heart and the resistance to blood flow of the capillaries. This **hydrostatic pressure** forces the fluid part of the blood through the capillary walls into the spaces between the cells.
- This outward flow is opposed by the reduced **water potential** of the blood, created by the presence of the plasma proteins.
- The hydrostatic pressure of the blood is greater than the osmotic forces so there is a net flow of water out of the blood. (As water passes through the capillary wall it carries solutes with it.)
- At the arterial end of the capillary bed the diffusion gradient for solutes such as glucose, oxygen and ions favours movement from the capillaries to the tissue fluid. This is because these substances are being used during cell metabolism.
- At the venous end of the capillary bed the blood pressure is lower and water passes into the capillaries by osmosis. The reduced water potential of the blood created by the presence of the plasma proteins causes a net inflow of water.
- At the venous end tissue fluid picks up CO_2 and other excretory substances. Some of this fluid passes back into the capillaries, but some drains into the lymphatic system and is returned eventually to the venous system via the thoracic duct, which empties into a vein near the heart.

(Other functions of the lymphatic system are studied at A2 and the following is for information only.) Lymph is the tissue fluid that drains into blind-ending lymphatic capillaries amongst the tissues. The lymph moves through vessels by contractions of the muscles through which the vessels pass. Lymph glands and lymph nodes associated with the lymph vessels play an important role in the formation of lymphocytes and the prevention of infection.)

Transport in plants

Roots absorb water from the soil and this water needs to be transported some distance to the leaves where it is used in the process of photosynthesis. In turn, the sugar produced must be transported to where it is needed. Plants have evolved two distinct systems of tubes, the xylem to transport water and phloem to transport sugars.

Vascular tissues

The vascular tissues are made up of:
- the **xylem tissue, which transports water and mineral salts from the roots to the leaves.**
- the phloem tissue, which transports the soluble products of photosynthesis (sucrose and amino acids) from the leaves to the other parts of the plant.

Structure of xylem

Xylem is made up of four different types of cells:
- **vessels -** the main conducting cells
- **tracheids -** also conduct water but are not as well adapted as vessels for this function
- **fibres** – they have no role in water transport but function in support
- **xylem parenchyma -** this acts as packing tissue.

Vessels and tracheids are dead cells and they form a system of tubes through which water can travel. The cells are dead because **lignin** has been deposited on the cellulose cell walls rendering them impermeable to water and solutes. These cells also provide mechanical strength and support to the plant.

Xylem distribution

The distribution of xylem tissue differs in primary stems, leaves and roots.
- In **stems** it occurs as part of the peripheral vascular bundles. This organisation gives flexible support but also resistance to bending strain.
- In **leaves** the arrangement of vascular tissues in the midrib and network of veins also gives flexible strength and resistance to tearing strains.
- In **roots** the central arrangement is ideal for resistance to vertical stresses (pull) and so helps the anchorage of the plant.

❑ ***Action***
1. Draw T.S. and L.S. diagrams of primary phloem.
2. Draw labelled diagrams of T.S. dicotyledon primary stem.
3. Draw labelled diagrams of T.S. dicotyledon primary root.

Water uptake by the roots

The large quantities of water lost through the stomata of the leaves by the process of transpiration must be replaced with water from the soil. The region of greatest uptake is the root hair zone where the surface area of the root is enormously increased by the presence of root hairs.

The soil water contains a very weak solution of mineral salts and so has a high water potential. The vacuole of the root hair cell contains a strong solution of dissolved substances and has a low water potential. Water passes into the root hair cell down a water potential (WP) gradient from a high WP to a low WP by osmosis. Water can travel across the cells of the cortex of the root along three pathways:
- the **apoplast** – through the cell wall
- the **symplast** – through the cytoplasm and plasmodesmata
- the **vacuolar pathway** – from vacuole to vacuole.

However, it is considered that the two main pathways are the symplast and apoplast pathways. Most of the water probably follows the latter pathway as this is the faster of the two.

The xylem tissue is found in the centre of the root and is surrounded by a single layer of cells called the **endodermis**. The cell walls of the endodermis are impregnated with a waxy material called suberin. This forms a distinctive band known as the **Casparian strip**. The suberin is waterproof and the Casparian strip prevents the use of the apoplast pathway. The only way that water can pass across the endodermis to the xylem is along the symplast pathway. There is some evidence that salts may then be actively secreted into the vascular tissue from the endodermal cells. This makes the WP in the xylem more negative, causing water to be drawn in from the endodermis, so promoting the movement of water into the xylem from the cortex. The water potential gradient produced creates a force known as the **root pressure.**

❑ ***Action*** Draw an endodermal cell showing the casparian strip.

The uptake of minerals

Generally, minerals are taken up by the root hairs by **active transport** from the soil solution.
Once absorbed the mineral ions may move along the apoplast pathway carried in solution by the water being pulled up the plant in the transpiration stream. When minerals reach the endodermis the Casparian strip prevents further movement along the cell walls. The ions enter the cytoplasm of the cell from where they diffuse or are actively transported into the xylem. For example, nitrogen usually enters the plant as nitrate ions/ammonium ions which diffuse along the concentration gradient into the apoplast stream but enter the symplast by active transport against the concentration gradient and then flow via plasmodesmata in the cytoplasmic stream.
At the endodermis, ions must be actively taken up to by-pass the Casparian band which allows the plant to selectively take up the ions at this point.

The movement of water from the root to the leaf

- Water travels in the xylem up through the stem to the leaves, where most of it evaporates from the internal leaf surface and passes out, as water vapour, into the atmosphere.
- The transpiration of water from the leaves draws water across the leaf from the xylem tissue along three same pathways as in the root.
- As water molecules leave xylem cells in the leaf, they pull up other water molecules. This pulling effect is known as the **transpiration pull** and is possible because of the large **cohesive** forces between water molecules and the **adhesive** forces which exist between the water molecules and the hydrophilic lining of the vessels. These two forces combine to maintain the column of water in the xylem.
- The theory of the mechanism by which water moves up the xylem is known as the **Cohesion-Tension theory.**
- **Capillarity** is another force that may contribute to the rise of water in the xylem. Water rises up narrow tubes by capillary action but this force is probably of more relevance in small plants than large trees.

❑ ***Action*** List the forces involved in the transport of water from the root to the leaf.

Transpiration

A major problem confronting all terrestrial organisms is how to avoid desiccation. Land plants in particular are continually losing water vapour to the atmosphere. In fact, about 99% of the water absorbed by the plant can be lost by evaporation through the leaves. This evaporation of water from inside the leaves through the stomata to the atmosphere is known as transpiration and gives rise to the **transpiration stream**. All plants have to balance water uptake with water loss. If a plant loses more water than it absorbs it wilts. If a plant loses an excessive amount of water it reaches a point where it cannot regain its turgor and it dies.

Plants face a dilemma. The stomata need to be open during the day to allow the exchange of gases between the tissues of the leaf and the atmosphere. However, the presence of pores in the leaves means that valuable water is lost from the plant. Most of the water is lost through the stomata although around 5% of the total water vapour loss can occur through the leaf epidermis. This loss is normally reduced due to the presence of the waxy cuticle on the surface of the leaves.

The rate at which water is lost from plants is called the **transpiration rate** and is dependent on external factors such as **temperature**, **humidity** and **air movement**. Any factor that increases the water potential gradient between the water vapour in the leaf and the surrounding atmosphere increases the rate of transpiration.

- **Temperature** – a rise in temperature provides additional kinetic energy for the movement of water molecules. This additional energy accelerates the rate of evaporation of water from the walls of the mesophyll cells and, if the stomata are open, speeds up the rate of diffusion of water vapour into the surrounding atmosphere. The water potential of the atmosphere becomes lower as its temperature is raised and it can hold more moisture.
- **Humidity** – the air inside a leaf is saturated with water vapour but the humidity of the atmosphere surrounding a leaf varies, with values exceeding 70% being rare in Britain. Thus the water potential gradient between leaf and atmosphere is always great and when the stomata are open water vapour rapidly diffuses from the leaf.
- **Air movement** – transpiration in still air results in the accumulation of a layer of saturated air at the surfaces of leaves. This offers considerable resistance to the diffusion of water vapour through stomata and thus reduces the rate of transpiration. Movement of the surrounding air reduces the thickness of the layer of saturated air and results in increased transpiration.
- **Light intensity** also affects transpiration by controlling the degree of stomatal opening.

In reality, these **factors** do not act independently but **interact** with each other, e.g. more water is lost on a dry, windy day than on a humid, still day. This is because the sub-stomatal chamber has a high water potential as the walls of the spongy mesophyll cells are saturated with water. The water evaporates from the walls and moves down a gradient of water potential from the plant to the atmosphere which has a low percentage relative humidity; the wind having reduced the thickness of the layer of saturated air at the leaf surface.

The rate of transpiration is measured using a **potometer.**
It actually measures the rate of water absorption but if the cells of the plant are fully turgid the rate of water absorption and the rate of transpiration are the same.

❑ ***Action*** Draw graphs showing how environmental factors affect the rate of transpiration.

Mesophytes, xerophytes and hydrophytes

Plants can be classified on the basis of structure, in relation to the prevailing water supply, into three groups. Hydrophytes (water plants), xerophytes (plants that live in conditions where water is scarce) and mesophytes (plants living in conditions of adequate water supplies). Most land plants growing in temperate regions belong to the latter category. Usually the water that they lose by transpiration is readily replaced by uptake from the soil, so they do not require any special means of conserving water. If such a plant loses too much water the plant wilts and the leaves droop. The leaf surface area is reduced and photosynthesis becomes less efficient.

- **Hydrophytes** grow submerged or partially submerged in water. An example is the water lily, which is rooted to the mud at the bottom of a pond and has floating leaves on the surface of the water. Hydrophytes are adapted as follows:

✓ As water is a supportive medium they have little or no lignified support tissues.
✓ Surrounded by water there is little need for transport tissue, so xylem is poorly developed.
✓ Leaves have little or no cuticle.
✓ Stomata are found on the upper surface of the leaves.
✓ Stems and leaves have large air spaces, forming a reservoir of oxygen and carbon dioxide. These gases also provide buoyancy to the plant tissues when submerged.

❑ ***Action*** Draw a plan diagram of the Water Lily T.S. leaf.

- **Xerophytes** are plants which show xeromorphic adaptations. These plants have adapted to living in conditions of low water availability and have developed modified structures to prevent excessive water loss. They may live in hot, dry desert regions; cold regions where the soil water is frozen for much of the year; exposed, windy locations.
  Ammophila arenaria (**marram grass**) is an example of a xerophyte which colonises sand dunes around the coast. The sand dune habitat makes it difficult for a mesophyte to survive there because there is no soil, rapid drainage of rain water occurs, there are high wind speeds, salt spray and a lack of shade from the sun.

❑ ***Action*** Draw, label and annotate diagram of T.S. leaf of marram grass.

Marram grass shows the following modifications:

✓ **Rolled leaves** - large thin-walled epidermal cells at the bases of the grooves shrink when they lose water from excessive transpiration, causing the leaf to roll onto itself. This has the effect of reducing the leaf area from which transpiration can occur.
✓ **Sunken stomata** - stomata are found in grooves on the inner side of the leaf. They are located in pits or depressions so that humid air is trapped outside the stomata. This reduces the water potential gradient between the leaf and the atmosphere and so reduces the rate of diffusion of water.
✓ **Hairs –** stiff, interlocking hairs trap water vapour and reduce the water potential gradient.
✓ **Thick cuticle -** the cuticle is a waxy covering over the leaf surface which reduces water loss. The thicker this cuticle the lower the rate of cuticular transpiration.

Additional xerophytic adaptations

✓ The development of thick walled sclerenchyma tissue in the leaf of Hakea prevents it from collapsing in times of drought and so maintains the surface area for photosynthesis.
✓ Cacti have succulent stems for water storage and leaves reduced to spines. Many cacti also have the ability to close their stomata during daylight.
✓ Pine trees have needle-like leaves. This reduces the surface area over which water is lost.

- **Mesophytes** flourish in habitats with adequate water supply. Most plants of temperate regions are mesophytes and, most importantly, most of our crop plants are mesophytes. They are adapted to grow best in well-drained soils and moderately dry air. Mesophytes lose a lot of water but excessive loss is prevented by closure of the stomata. Water uptake during the night makes good the water lost during the day.
  Mesophytes need to survive unfavourable times of the year, particularly when the ground is frozen.

✓ Many trees and shrubs shed their leaves before the onset of winter.
✓ The aerial parts of many non-woody plants die off as a result of frost or cold winds but their underground organs survive e.g. bulbs, corms.
✓ Most annual mesophytes (plants that flower, produce seed and die in the same year) survive the winter as dormant seeds.

Translocation

The products of photosynthesis are transported in the phloem, away from the site of synthesis in the leaves (the 'source'), to all the other parts of the plant where they are used for growth or storage (the 'sink'). In plants the transport of the soluble organic materials, sucrose and amino acids, is known as translocation.

Structure of phloem

Phloem is a living tissue and consists of four types of cells:

- **sieve tubes**
- **companion cells**
- **phloem fibres**
- **phloem parenchyma.**

The sieve tubes are the only components of phloem obviously adapted for the longitudinal flow of material. They are formed from cells called sieve elements placed end to end. The end walls do not break down but are perforated by pores. These areas are known as **sieve plates**. Cytoplasmic filaments containing phloem protein extend from one sieve cell to the next through the pores in the sieve plate. The sieve tubes do not possess a nucleus and during their development most of the other cell organelles disintegrate. Each sieve tube element is closely associated with at least one companion cell, which have dense cytoplasm, large centrally placed nuclei, many mitochondria, and they are connected to the sieve tube element by **plasmodesmata**.

❑ ***Action***

1. Draw T.S. and L.S. of primary phloem.
2. Observe the structure of phloem as seen under the electron microscope.

Transport in the phloem

There is a lot of experimental evidence to suggest that the phloem is the tissue concerned with the translocation of organic substances such as sucrose and amino acids. Early evidence was obtained from **ringing experiments** where cylinders of outer bark tissue (removing the phloem) were removed from woody stems and the contents of the phloem above and below the cylinder were later analysed. More recently, the technique of radioactive tracing with labelled metabolites and direct sampling of the contents of individual sieve tubes, using the mouth parts of aphids as micro-pipettes, have been used. Aphids have hollow, needle-like mouthparts (stylet) which they insert into sieve tubes to feed on the sap. To sample the phloem sap the aphid is anaesthetised and the stylet is cut off, leaving it attached to the plant. As the sap is under pressure it exudes from the very fine tube and can be collected and analysed. These experiments have also enabled scientists to demonstrate that translocation is a rapid process, much too rapid to be explained by diffusion.

Radioisotope labelling is a technique where carbon dioxide labelled with radioactive carbon is supplied to an illuminated plant leaf. The radioactive carbon is fixed in the sugar produced in photosynthesis and its translocation to other parts of the plant can be traced using autoradiography. The 'source' leaf and 'sink' tissues are placed firmly on photographic film in the dark for 24 hours. When the film is developed, the presence of radioactivity in parts of the tissue show up as 'fogging' of the negatives. The technique shows that the sugar is transported in both an upward and downward direction since the radioactivity is shown in the aerial parts of the plant as well as the roots.
An autoradiograph of a transverse section of a stem of a treated plant shows fogging only where phloem was in contact with the film.

Theories of translocation

The observed rate of flow is much too rapid for diffusion to be the cause. It has been observed that sucrose is translocated in sieve tubes at a rate of 25 – 100 cm h^{-1} compared with only 0.2 mm day^{-1} possible by diffusion.

The main theory put forward to explain the transport of organic solutes is known as the **mass flow hypothesis** (1937). This theory suggests that there is a passive mass flow of sugars from the phloem of the leaf where there is the highest concentration (the source) to other areas, such as growing tissues, where there is a lower concentration (the sink).
(The details of the mechanism of this theory are not required.)

Arguments against the theory include:

- ✓ It does not explain the existence of the sieve plates which seemingly act as a series of barriers impeding flow.
- ✓ Sucrose and amino acids have been observed to move at different rates and in different directions in the same tissue.
- ✓ Phloem tissue has a relatively high rate of oxygen consumption and translocation is slowed down or stopped altogether if respiratory poisons such as potassium cyanide enter the phloem.
- ✓ The companion cells contain numerous mitochondria and so release energy but the mass flow hypothesis does not suggest a role for the companion cells.

Recent theories put forward suggest:

- ✓ An active process may be involved.
- ✓ The observation of streaming in the cytoplasm of individual sieve tubes could be responsible for bi-directional movements along individual sieve tubes, providing there was some mechanism to transport solutes across the sieve plates.
- ✓ Some scientists have observed protein filaments passing through the sieve pores and suggest that different solutes are transported along different filaments.

At present scientists do not have a complete understanding of the mechanism of translocation and a great deal of debate continues.

(You are not required to give details of any of the theories.)

2.4 Reproductive strategies

The life cycle of an organism may be defined as the sequence of changes through which it passes during its life, from its origin in reproduction until its death. For a species to survive it must produce new individuals. Reproduction is the ability to produce other individuals of the same species and is a fundamental characteristic of living things. New individuals may be produced by either asexual or sexual reproduction, or in some species by both methods. In animals asexual reproduction is far less common than it is in plants, protoctists and prokaryotes.

Asexual and sexual reproduction

In plants and animals reproduction is achieved in two ways.
Asexual reproduction – this method rapidly produces large numbers of individuals having an identical genetic composition. A group of genetically identical offspring produced by this method is called a **clone.** Examples of asexual methods in animals include binary fission and budding. Plant examples are bulbs e.g. daffodil, runners e.g. strawberry, tubers e.g. potatoe.
Sexual reproduction – this method usually involves two parents, is less rapid than asexual and produces offspring that are genetically different. The fusion of haploid gametes is always involved.

Advantages and disadvantages

In asexual reproduction lack of variety is a disadvantage in adapting to environmental change but its main advantage is that if an individual has a genetic makeup suited to a particular set of conditions, large numbers of this successful type may be built up.
Advantages of sexual reproduction:

- ✓ There is an increase in genetic variety so enabling a species to adapt to environmental change.
- ✓ It allows the development of a resistant stage in the life cycle which enables the species to withstand adverse conditions.
- ✓ The formation of spores, seeds, and larvae enables the dispersal of offspring. This reduces intra-specific competition and so enables genetic variety to develop as required.

Mutations, although rare, help to create a little variety in asexual reproduction. Mutations arise more frequently (but still rarely) during sexual reproduction because of the greater complexity of the process.

The production of gametes

Living organisms have **diploid** body cells and **haploid** sex cells or **gametes.**
Body cells with the full chromosome number are made by **mitosis**.
Haploid cells with half the chromosome number are produced by **meiosis**.
At fertilisation the haploid sperm fuses with a haploid egg to produce a diploid fertilised egg.
The zygote formed then divides many times by mitosis to grow into a new individual.
Males and females usually produce different sized gametes. The **male** gamete is small and extremely **motile** and the **female** gamete is large and sedentary, normally due to the presence of **stored food**. Mammalian eggs differ in that they contain very little stored food and instead the materials for development are obtained from the maternal blood supply through the placenta.

External and internal fertilisation

- Many aquatic organisms discharge their gametes directly into the sea or freshwater. As the gametes are quickly dispersed by the water, there is a strong possibility that many eggs will not encounter sperm. In animals this sort of fertilisation is known as **external** fertilisation. There is a considerable wastage and so large numbers of gametes of both sexes have to be produced. In the frog the joining of sperm and egg is facilitated by a sexual coupling. When the eggs are laid by the female, the male immediately releases seminal fluid over them. While amphibians in general can move efficiently on land, many must return to water to breed and fertilisation is external.

- In most terrestrial animals, however, fertilisation occurs inside the body of the female and this is called **internal** fertilisation. Generally, this requires the use of some kind of intromittent organ to introduce the sperm into the female's body.

  Internal fertilisation has several advantages;

1. There is less chance of gametes being wasted.
2. It allows the male gamete to become independent of the need for water for movement.
3. The fertilised egg can be enclosed within a protective covering before it leaves the female's body. This is what happens in animals that lay eggs. Some animals take this idea further and the embryos develop within the female parent and derive nourishment from her. This reaches its greatest development in those mammals that nourish their developing young before birth by means of a placenta.

Development of the zygote

In many animals the fertilised egg or zygote undergoes its development outside the body of the parent. The developing offspring are easy prey for predators and provide food for other species. Many eggs are produced to ensure that at least a few survive. In insects, although fertilisation is internal, the fertilised eggs are normally laid on a suitable food source and the embryo develops outside the body. Internal fertilisation ensures that all the sperm are deposited in the female's reproductive tract.
The gradual adaptation to life on land includes the evolution of an amniote egg in reptiles and birds. The egg has a fluid filled cavity surrounded by a membrane outside which is a protective shell which encloses the embryo within the yolk sac. Birds incubate eggs and the embryo completes its development outside the mother's body.
In mammals the young are retained for a considerable time in the mother's womb or uterus but there is no shell. The embryo is nourished there from the mother's blood supply via the placenta. The young are born in a relatively advanced state of development.

Parental care

A stated previously, many animal species reproduce by forming and laying fertilised eggs that are left to develop unattended. There is little or no parental care. Others provide some sort of parental care e.g. the male stickleback looks after the fertilised eggs in a defended territory and fans them to provide oxygen until they hatch and swim away. Pronounced parental care is typical of most species of birds and mammals. It includes the provision of shelter from unfavourable environmental conditions, feeding, protection from predators, and in some species the training of their offspring as they prepare for adult life. Generally, the more parental care provided the fewer the number of offspring produced. Certain types of fish produce over 100 million eggs at a spawning. At the other extreme certain mammals, including humans, usually produce only one offspring at a time.

Unisexual and hermaphrodite

In most animals male and female sex organs are contained in different individuals i.e. the individuals are unisexual. In most flowering plants male and female gametes are produced in each individual i.e. the individuals are hemaphrodite. This has a biological advantage in that every individual is capable of forming fertilised eggs. The disadvantage is the possibility of self-fertilisation leading to inbreeding i.e. the production of offspring with reduced genetic variability. Plants may not possess intromittent organs as such but special techniques have evolved for transferring the male gametes to the egg cells after which early development of the embryo takes place within the body of the parent plant.
(No details of flower structure or methods of pollination are required.)

The success of insects

Insects are a very successful group of mainly terrestrial animals (although the larvae of several species are aquatic). In species diversity, insects outnumber all other forms of life combined, with over 1 million different species. As they are so numerous, diverse and widespread, they affect the lives of all other terrestrial organisms particularly humans. Humans depend on insects such as bees to pollinate crops but insects are also carriers of disease including malaria and African sleeping sickness. Insects also compete with humans for food and billions of pounds are spent annually on insecticides.

In insects, during the development of the zygote, an intermediate form, referred to either as the **nymph** or the **larva,** is formed. These are juvenile forms that develop from the egg stage in the life cycle. Insects have a hard outer skeleton (exoskeleton) and in order to grow they have to shed the skin or moult. They do this several times during their development. Grasshoppers and locusts are examples of insects that develop from nymphs. They are said to undergo an **incomplete metamorphosis** where the young nymph, which resembles the adult, hatches from the fertilised egg and goes through a series of moults until it reaches full size.

The majority of insect species develop from a larval stage which is quite different from the adult. The process involves considerable changes and is called a **complete metamorphosis**. These insects, such as the butterfly and housefly, also have an additional stage called the **pupa** or chrysalis. The larva hatches from the egg and is specialised for feeding and growing. The larva undergoes a period of change within the pupa and emerges as the adult that is specialised for dispersal and reproduction.

Plants

Simple plant forms such as algae e.g. seaweeds, are confined to an aquatic environment for at least part, if not all, of their lives. Other plant groups such as mosses and ferns are confined to damp areas as the male gametes require a surface film of water in which to swim to the egg. Are you surprised that the less advanced plant life forms possess motile gametes like animals? Like the successful land animals, the conifers and flowering plants became independent of water for their reproduction and so were able to colonise the land.

Flowering plants are well suited to life on land both in terms of their morphology e.g. efficient water carrying xylem vessels, and in their reproduction. They developed male pollen grains with a hard coat that could withstand desiccation. This enabled their transfer to the stigma of the female part of the plant without being affected by environmental conditions. Plants such as grasses have small, green inconspicuous flowers and the pollen is carried by wind. In plants with brightly coloured flowers and scent for attraction the pollen is carried to the stigma by insects.. The male gametes travel through the tissue of the female part and to the egg by means of a pollen tube. This means that sexual reproduction no longer depends on gametes having to travel through a film of water to reach the egg cell. Following fertilisation, the fertilised egg develops into a seed containing a food store.

The flowering plants are the most successful of all terrestrial plants. There are more than 300,000 species. They are found in every type of habitat. A key feature of their success is their relationship with animals e.g. plants attract animals, particularly insects, to their flowers in order to feed and so exploit their mobility for pollination and seed dispersal. Another important development was the enclosure of the eggs in an **ovary** and the evolution of the **seed**. Seeds that result from fertilisation contain food reserves and have a resistant coat and so are able to withstand adverse conditions.

Why did the flowering plants become so successful?

1. The interval between flower production and the setting of seed is usually a matter of weeks.
2. The production of the seed with a food store enables the embryo to develop until leaves are produced above ground and are able to photosynthesise. The seed also protects the embryo from desiccation and other hazards.
3. Generally, leaves are deciduous and succulent and decay rapidly on falling to the ground. This enables humus to be produced and as a consequence the rapid recycling of ions for reuse by plants.

2.5 Adaptations for nutrition

Types of nutrition

Nutrition is the process by which organisms obtain **energy** to maintain life functions and **matter** to create and maintain structure. These are obtained from nutrients.
A**utotrophic** organisms such as green plants use the simple organic materials carbon dioxide and water to manufacture energy - containing complex organic compounds, whereas **heterotrophic** organisms consume complex organic food material.

Autotrophic nutrition

Living organisms that can make their own food are called autotrophs.
They provide food for all other life forms and so they are also known as producers.
There are two types of autotrophic nutrition:

- **Photosynthesis** is the process by which green plants build up complex organic molecules such as sugars, from carbon dioxide and water. The source of energy for this process comes from sunlight which is absorbed by chlorophyll and related pigments. Algae and certain types of bacteria can also photosynthesise using energy from sunlight.
- **Chemosynthesis** is a process carried out by autotrophic bacteria. They use the energy derived from special methods of respiration to synthesise organic food.

Heterotrophic nutrition

Heterotrophs cannot make their own organic food. They have to consume complex organic food material produced by autotrophs. Since they eat or consume ready-made food they are known as consumers. All animals are consumers and are dependent on producers for food. Heterotrophs include animals, fungi, some types of protoctists and bacteria.

There are a number of different types of heterotrophic nutrition.

- **Holozoic feeders**

This includes nearly all animals. They take their food into their bodies and break it down by the process of digestion. Most carry out this process inside the body within a specialised digestive system. The digested material is then absorbed into the body tissues and used by the body cells. Animals that feed solely on plant material are termed **herbivores**, those that feed on other animals are **carnivores**, and **detritivores** are animals that feed on dead and decaying material.

- **Saprophytes**

This group is also known as saprobionts, and include all fungi and some bacteria. They feed on dead or decaying matter and do not have a specialised digestive system. They feed by secreting enzymes such as proteases, amylases, lipases and cellulases onto the food material **outside** the body and then absorb the soluble products across the cell membrane by diffusion. This is known as **extracellular digestion**. Microscopic saprophytes are called **decomposers** and their activities are important in the decomposition of leaf litter and the recycling of valuable nutrients, such as nitrogen.

- **Parasites**

A parasite is an organism that feeds on another living organism, referred to as the host. Some parasites live in the body of the host while others live on the surface. The host always suffers harm to some degree and often death. Parasites are considered to be very highly specialised organisms and show considerable adaptations to their particular way of life. Examples of parasites are the tapeworm, potato blight, caused by a fungus, and *Plasmodium,* the malarial parasite.

- **Mutualism**

This is also known as symbiosis and involves a close association between members of two different species, but in this case both derive some benefit from the relationship e.g. the digestion of cellulose by microorganisms in the gut of a herbivore. Cows and sheep feed mainly on grass, a high proportion of which is made up of cellulose cell walls. Like other herbivores cows and sheep do not secrete the enzyme cellulase and so cannot digest cellulose. Instead they have mutualistic bacteria which live in a specific part of the specialised stomach (the rumen). These bacteria produce the enzymes for them and in return the bacteria gain other digestive products and suitable conditions for growth.

Structure of the gut

Hterotrophs cannot synthesise their major food requirements and so are dependent on a source of ready-made food. This food is needed as a source of energy for activities such as movement and for the synthesis of body tissues. The organic molecules must be broken down by digestion and absorbed into the body tissues from the digestive system before utilisation in the body cells. Digestion and absorption take place in the gut, which is a long, hollow, muscular tube. The gut is organised to allow movement of its contents in one direction only. In simple organisms which feed on only one type of food the gut is undifferentiated. However, in more advanced organisms with a varied diet the gut is divided into various parts along its length and each part is specialised to carry out particular steps in the processes of mechanical and chemical digestion as well as absorption.

The functions of the digestive system

In the human the main regions of the gut are the mouth, oesophagus (gullet), stomach, small intestine (duodenum, ileum), large intestine and anus. The food is processed as it passes along the various regions of the gut. It is propelled along the gut by the process of peristalsis.

- **Ingestion** is the taking in of food into the body through the mouth.
- **Digestion** is the breakdown of large, insoluble food molecules into simple, soluble molecules by means of enzymes. **Mechanical** digestion in humans is achieved by the cutting and/or crushing action of the teeth followed by the rhythmical contractions of the gut. The gut wall, particularly the stomach, has layers of muscle to fulfil this function. These are responsible for mixing the food and pushing it along the gut. The physical action also has an important role as it increases the surface area over which enzymes can act. The **chemical** action of digestion is achieved through the secretion of digestive enzymes. The enzymes are secreted either by:

✓ glands situated outside the gut, e.g. salivary glands and pancreas
✓ glands located in the gut wall itself

- **Absorption** is the passage of digested food through the gut wall into the blood.
- **Egestion** is the elimination from the body of food that cannot be digested e.g. cellulose cell walls of plants.

❑ ***Action*** Draw a labelled diagram of the human alimentary canal and its associated organs. (You should label the following - buccal cavity, tongue, salivary glands, oesophagus, stomach, duodenum, ileum, colon, rectum, anus and associated organs; liver and pancreas.)

The structure of the mammalian gut

Throughout its length from the mouth to the anus the gut wall consists of four tissue layers surrounding a cavity (lumen) of the gut.

- The outer serosa consists of a layer of tough connective tissue that protects the wall of the gut and reduces friction from other organs in the abdomen as the gut moves during the digestive process.
- The **muscle** layer consists of two layers of muscle running in different directions:

✓ the inner circular muscle
✓ the outer longitudinal muscle.

Collectively these muscles cause waves of muscular contractions, **peristalsis**, which propels food along the gut. Behind the ball of food the circular muscles contract and the longitudinal muscles relax, thus helping move the food along.

❑ ***Action*** Draw a labelled diagram to illustrate peristalsis.

• The **sub-mucosa** consists of connective tissue containing blood and lymph vessels to take away absorbed food products as well as nerves that co-ordinate the muscular contractions involved in the process of peristalsis.

• The **mucosa** is the innermost layer and lines the wall of the gut. It secretes mucus which lubricates and protects the mucosa. In some regions of the gut this layer secretes digestive juices in others it absorbs digested food.

❑ ***Action*** Histological structure of the ileum.
(Draw the ileum and label the following - serosa, longitudinal muscle layer, circular Muscle layer, submuscosa, muscularis mucosa, mucus membrane, goblet cells, lacteal, villi and muscle.)

Glands in the gut

Glands produce a large amount of secretions, some of which contain digestive enzymes.
The glands of the gut are of three types:

- large glands found outside the gut with secretions passing through tubes or ducts into the gut cavity
  - ✓ salivary gland which secretes saliva into the mouth
  - ✓ liver which secretes bile into the duodenum
  - ✓ pancreas which secretes pancreatic juice into the duodenum
- glands in the form of cells in the submucosa
  - ✓ glands which secrete mucus into the duodenum
- glands in the form of cells in the mucosa
  - ✓ gastric glands in the stomach wall which secrete gastric juice into the stomach
  - ✓ glands found at the base of the villus in the small intestine which secrete enzymes into the small intestine.

Digestion

As humans have a varied diet the gut is divided into various parts along its length. Each region is specialised to carry out particular steps in the processes of mechanical and chemical digestion as well as absorption. At this point you should review work on 'Biological compounds and enzymes' in Unit BY1.

Digestion

The absorption of nutrients by the gut epithelial cells is only possible if the large molecules, carbohydrates, fats and proteins are first broken down or digested into smaller products by means of enzymes. These digestive enzymes are known as **hydrolases** and catalyse the hydrolysis of a substrate by the addition of a molecule of water. Different enzymes are required to carry out the digestion of the different food substrates and usually more than one type of enzyme is needed for the complete digestion of a particular food.

During digestion:

- **Carbohydrates** (polysaccharides) are first broken down into disaccharides and then into monosaccharides. The enzyme **amylase** hydrolyses starch to the disaccharide maltose but another enzyme **maltase** is required to break down the maltose to the monosaccharide, **glucose.**
- **Proteins** are broken down into polypeptides, then dipeptides, and finally into **amino acids**.
  The general name given to the protein digesting enzymes is **peptidase.**
  Proteins are extremely large molecules so **endopeptidases** hydrolyse peptide bonds within the protein molecule and **exopeptidases** hydolyse peptide bonds at the ends of these shorter polypeptides.
- **Fats** are broken down to **fatty acids and glycerol** by just one enzyme, **lipase.**

Different enzymes have different pH optima so the different enzymes function in particular areas of the gut. The following section deals with the functions of these different regional specialisations.

Regional specialisations

The mouth

Mechanical digestion begins in the mouth when food is chewed using the teeth. The food is also mixed with saliva from the salivary glands. Saliva is a watery secretion, containing mucus and salivary amylase, together with some mineral ions which help to keep the pH in the mouth slightly alkaline, the optimum pH for amylase. Saliva is important for lubricating the food before it is swallowed. **Amylase** breaks down **starch** to **maltose**. After chewing the ball of food is swallowed and mucus lubricates its passage down the oesophagus.

The stomach

Food enters the stomach and is kept there by the contraction of two rings of muscles, one at the stomach entrance and one at the junction with the duodenum. Food may stay in the stomach for up to four hours and during this time the muscles of the stomach wall contract rhythmically and mix up the food with gastric juice secreted by glands in the stomach wall. Gastric juice contains acid that gives the stomach contents a pH of 2.0. As well as providing the optimum pH for the enzymes, the acid kills most bacteria in the food. Gastric juice also contains **peptidase** enzymes which hydrolyse the **protein** to **polypeptides**. Mucus is important in forming a lining to protect the stomach wall from the enzymes and acid as well as assisting in the movement of food within the stomach.

The small intestine

The small intestine is divided into two regions, the duodenum and the ileum.
Relaxation of the muscle at the base of the stomach allows small amounts of the partially digested food into the duodenum a little at a time. The duodenum makes up the first 20cm of the small intestine and receives secretions from both the liver and the pancreas.

- **Bile** is produced in the liver and stored in the gall bladder from where it passes into the duodenum via the bile duct. It contains no enzymes but the bile salts are important in **emulsifying** the **lipids** present in the food. Emulsification is achieved by lowering the surface tension of the lipids, causing large globules to break up into tiny droplets. This enables the action of the enzyme **lipase** to be more efficient as the lipid droplets now have a much larger surface area. Bile also helps to neutralise the acidity of the food as it comes from the stomach.
- The **pancreatic juice** is secreted from the exocrine glands in the pancreas and enters the duodenum through the pancreatic duct. It contains a number of different enzymes:

✓ **endopeptidases** which hydrolyse **protein** to **peptides**.
✓ **amylase**, which breaks down any remaining **starch** to **maltose.**
✓ **lipase,** which hydrolyse **lipids** into **fatty acids and glycerol**.

The walls of the duodenum contain glands that secrete an alkaline juice and mucus. The alkaline juice helps to keep the contents of the small intestine at the correct pH for enzyme action, and the mucus is for lubrication and protection.

Enzymes secreted by cells at the tips of the villi complete digestion.
✓ **maltase** hydrolyses **maltose** into two **glucose** molecules.
✓ **endopeptidases** and **exopeptidases** complete the digestion of polypeptides to amino acids.

The end products of carbohydrate digestion are all monosacharides. The final stage of carbohydrate digestion is intracellular, as disaccharides are absorbed by the plasma membrane of the epithelial cells before being broken down into monosaccharides.

Absorption

The ileum is well adapted for absorption:

- In humans the ileum is very long.
- The lining is folded to give a large surface area compared to a smooth tube.
- On the folds are numerous finger-like projections called villi.
- On the surface of the villi are epithelial cells with microscopic projections called microvilli. These increase the surface area of the cell membrane of the epithelial cells for absorption.

Absorption follows digestion and takes place mainly in the small intestine. Because energy is required for active absorption, the epithelial cells also contain large numbers of mitochondria.

❑ ***Action*** Draw a detailed section through a villus and show the regions of absorption of the different products of digestion.

- **Glucose and amino acids** are absorbed across the epithelium of the villi by a combination of diffusion and active transport. They pass into the capillary network that supplies each villus. The blood from the venules, which contains the dissolved food, eventually reaches the hepatic portal vein and and is carried to the liver.
- **Fatty acids and glycerol** are passed into the lacteal. This is a blindly ending lymph capillary found in the centre of each villus. Fatty acids and glycerol are transported in the lymphatic system which ultimately opens into the blood stream at the thoracic duct.

To summarise, the following methods of transport occur:
Fatty acids, glycerol and most vitamins pass through the membrane of the epithelial cells by diffusion. However, the glucose, amino acids and dipeptides require energy in the form of ATP for absorption by active uptake. Dipeptides are then digested intracellularly into simple amino acids. Glucose and amino acids then diffuse from the epithelial cell into the blood.

The large intestine

The large intestine is about 1.5 metres long and is divided into the caecum, the appendix, the colon and the rectum. Water and mineral salts are absorbed from the colon along with vitamins secreted by microorganisms living in the colon. These bacteria are responsible for making vitamin K and folic acid. By the time it reaches the rectum, indigestible food is in a semi-solid condition. It consists of residues of undigested cellulose, bacteria and sloughed cells and passes along the colon to be egested as faeces. This process is called defaecation.

Uses of the products of digestion

With the completion of the processes of digestion and absorption the soluble food products are carried in the bloodstream to the tissues for **assimilation** or to provide **energy**.

- Glucose is absorbed from the blood by cells, for energy release in respiration, and any excess is stored in fat cells.
- Amino acids are absorbed for protein synthesis, excess cannot be stored so are deaminated whereby the removed amino groups are converted to urea and the remainder to carbohydrate and stored.
- Lipids are used for membranes and hormones, excess are stored as fat.

Adaptations to different diets

Reptiles and amphibians swallow food whole immediately it is caught, but in mammals food is retained in the mouth whilst it is cut up and chewed. Mammals have a palate that separates the air path (nasal cavity) from the mouth. This allows food to be retained in the mouth rather than swallowed whole between breaths. The gut of a carnivore is short reflecting the ease with which protein is digested. However, the gut of a herbivore is long because digestion of plant material is difficult.
Since food is retained for cutting, crushing, grinding or shearing according to diet, mammals have evolved different types of teeth with each type being specialised for a different function.
Herbivores and carnivores have teeth specialised to suit their diets.

Dentition

Teeth are important in the mechanical digestion of food. Chewing food is important as it makes it easier to swallow and also increases the surface area for enzyme action.
Humans are omnivore, that is they eat they both plant and animal material. The teeth are not particularly specialised but having four different types of teeth to carry out different functions reflects a mixed diet. In total humans have 32 teeth and these are made up of:

- Eight chisel -shaped **incisors** at the front of the mouth for biting and cutting.
- Four pointed **canines** that function as incisors.
- Ten large, flat teeth on each side used for chewing. These are the **premolars** and **molars**.

Dentition in herbivores

Most of the meat and milk consumed by humans is obtained from cows and sheep. It is therefore important that we understand how they feed.
Plant food is a tough material and the teeth of herbivores are modified to ensure that it is thoroughly ground up before it is swallowed. A grazing herbivore, such as a cow or sheep, has incisors on the lower jaw only and cuts against a horny pad on the upper jaw. The canine teeth are indistinguishable from the incisors. A gap called the diastema separates the front teeth from the side teeth or premolars. The tongue operates in this gap moving the freshly-cut grass to the large grinding surfaces of the cheek teeth. The jaw moves in a circular grinding action in a horizontal plane. The cheek teeth interlock, like the letter W fitting into the letter M. With time the grinding surfaces become worn down, exposing the sharp-edged enamel ridges which further increase the efficiency of the grinding process. The teeth have open, unrestricted roots so that they can continue to grow throughout the life of the animal.

Dentition in carnivores

Carnivorous mammals, such as a tiger, have teeth adapted for catching and killing prey, cutting or crushing bones and for tearing meat. The sharp incisors grip and tear flesh from bone. The canine teeth are large, curved and pointed for seizing prey, for killing and also tearing flesh. The premolars and molars are for cutting and crushing. Carnivores have a pair of specialised cheek teeth, called carnassials, which slide past each other like the blades of gardening shears. The jaw muscles are well developed and powerful to enable the carnivore to grip the prey firmly and help in crushing bone. There is no side to side movement of the jaw, found only in herbivores, as this would lead to the jaw being dislocated when dealing with prey. The vertical jaw movement is greater than in herbivores allowing the jaw to open widely for capturing and killing prey.

- ❑ ***Action*** Draw labelled diagrams showing the dentition of a sheep and a cat.

Ruminants

Animals produce 35% of all the protein eaten by humans. Of this nearly half comes from **ruminants**, such as cows and sheep, that eat mainly grass and forage, a large proportion of which consists of cellulose cell walls. Ruminants have a specialised stomach or rumen in which mutualistic bacteria live. Mutualism or symbiosis involves a close association between members of two different species and where both organisms derive some benefit from the relationship. Herbivorous mammals such as cows and sheep lack the ability to produce cellulase enzymes and so cannot digest cellulose. A considerable proportion of plant material consists of cellulose cell walls. Certain herbivores, e.g. cows, have formed an association with cellulose-digesting bacteria which live in the gut of the cow. In this association the mammal acquires the products of cellulose digestion and the bacteria receive a constant supply of food and can grow in a suitable sheltered environment.

The cow provides a region of the gut for the bacteria to inhabit and in return the bacteria digest the cellulose for the cow. However, the region of the gut must be kept separate from the main digestive region so that:

✓ food can be kept there long enough for the bacteria to carry out the digestion of the cellulose.
✓ the bacteria are isolated from the mammal's own digestive juices so that they are in the optimum pH for their activities and they are not killed by extremes of pH.

Ruminants have a 'stomach' which is made up of four chambers. Three of the chambers are derived from the lower part of the oesophagus and one chamber is the true stomach.

Cellulose digestion takes place as follows:

- The grass is chopped by the teeth, mixed with saliva, and the cud formed is swallowed.
- In the rumen the first 'stomach', the cud is mixed with cellulose digesting bacteria to produce glucose. This is fermented to form organic acids, that are absorbed into the blood, and provides energy for the cow. The waste products are carbon dioxide and methane that are passed out.
- The cud passes to the next region before being regurgitated into the mouth and chewed again.
- The cud passes directly into the third 'stomach' where water is reabsorbed.
- The fourth and last 'stomach' functions like a 'normal' stomach and protein is digested.
- The digested food passes to the next region, the small intestine, where the products of digestion are absorbed.

The ruminants are more efficient in terms of nutrients extracted from cellulose than other herbivores. The rumen contains a greater variety of mutualistic organisms than the caecum and they are able to achieve a more complete breakdown of cellulose. Another advantage in ruminants is that the rumen is the first 'stomach' located before the main region of enzyme production so that when the bacteria die they pass through this region with the food and form an important source of protein.

❑ ***Action*** Compare the gut regions of a herbivore with that of a ruminant.

Parasitic nutrition

Parasites are organisms that live on or in another organism, called the host, and obtain nourishment at the expense of the host. Parasites therefore cause harm to some degree and often cause death. Many organisms are parasitised for at least part of their lives. Plants are parasitised by bacteria, fungi, viruses, nematodes and insects; animals are parasitised by bacteria, fungi, viruses, proctista, tapeworms, nematodes, insects and mites. Even bacteria are parasitised by viruses called bacteriopgages! The study of parasites is of economic importance as they cause disease in humans, crops and domesticated animals.

The pork tapeworm – a parasite of the gut

All animals have a struggle to survive, to avoid competition with others and to avoid being preyed upon by other animals. Parasites have become specialised and undergone considerable evolutionary changes in order to survive in the host. The gut parasite (*Taenia solium*) is a particularly good example. Imagine living in the gut of another animal! The tapeworm is ribbon-like and can be up to 10 metres long! It has a 'head' made up of muscle on which are suckers and hooks. Its body consists of a linear series of thin segments. The pork tapeworm has two hosts. The primary host is the human and the pig is the secondary host. The pig becomes infected if it feeds in drainage channels contaminated by human faeces. Humans are infected by eating undercooked infected pork.
(A knowledge of the life cycle is not required.)

Although the tapeworm lives in an immediate source of food it needs to survive the hostile conditions found in the gut.
The following are the problems that the gut parasite has to overcome in order to survive:

- It lives surrounded by digestive juices and mucus.
- Food, mixed with digestive juices, is in constant motion as it is churned about as well as being propelled along the length of the gut by peristaltic contractions of the muscular wall.
- It lives in extremes conditions of pH along the length of the gut.
- The immune system of the host.
- If the host dies then so does the parasite.

Adaptations for survival

In order to survive the tapeworm must:

- Have a means of penetrating the host
- Have a means of attachment to the host
- Protect itself against the immune responses of the host
- Develop only those organs that are essential for survival
- Produce many eggs
- Have an intermediate host
- Have resistant stages to overcome the period away from a host

They have evolved the following structural modifications for their parasitic way of life:

- Suckers and a double row of curved hooks for attachment to the wall of the gut.
- A body covering which protects them from the host's immune responses.
- A thick cuticle and the production of inhibitory substances on the surface of the segments to prevent their digestion by the host's enzymes.

- Because they live in a stable environment they do not need to move around and do not require a sensory system. This has led to the degeneration of unnecessary organs. They do have a simple excretory and nervous system but most of the body is concerned with reproduction.
- The tapeworm is very thin and has a large surface area to volume ratio. It is surrounded by digested food so it has a very simple digestive system and pre-digested food can be absorbed over the entire body surface.
- Because the gut could not accommodate two tapeworms each segment contains both male and female reproductive organs. Vast numbers of eggs are produced, with each mature segment containing up to 40,000 eggs. The mature segments pass out of the host's body with the faeces. The eggs have resistant shells and can survive until eaten by the secondary host. Further development can then take place and the embryos which hatch from the eggs move into the muscles of the pig and remain dormant until the meat of the pig is eaten by a human.

Harmful effects of the pork tapeworm.
The adult worms cause little discomfort but, if the eggs are eaten by humans, dormant embryos form cysts in various organs and damage the surrounding tissue. Adults can be treated with appropriate drugs. Public health measures and frequent inspection of meat are essential measures.

❑ ***Action*** Draw and label the 'head' of a tapeworm showing attached segments.